AUSTRALIAN GRASS PARRAKEETS

DEDICATION

We dedicate this book to our friend
and fellow aviculturist, the late
Herbert (Stud) Baker
who guided Stan through his
early years in aviculture.

AUSTRALIAN GRASS PARRAKEETS

(The Neophema)

Experiences in the Field and Aviary

by
STAN SINDEL
JAMES GILL

SINGIL PRESS PTY LTD
P.O. Box 9, Austral, NSW 2171
Australia

The National Library of Australia
Cataloguing-in-Publication entry:

Stan Sindel.

Australian grass parakeets, genus neophema
experiences in field and aviary.

Bibliography.

ISBN 0 9587727 4 6.

1. Budgerigar — Australia. 2. Aviculture — Australia. I. Gill, James
(James Harley). II. Title.

636.6864.

Printed in Australia by

SURREY BEATTY & SONS PTY LIMITED
43 Rickard Road, Chipping Norton, NSW 2170
Australia

ACKNOWLEDGEMENTS

We extend our sincere thanks to the multitude of people throughout Australia who helped us in various ways to compile the information contained within these pages.

Our particular thanks to all of the following people: Jack Stunnell, Stan Simmons, Kim and Roger Simmons, Russell McAllister, Bill Connor, Harvey Oliver, Gary Hall, Daryl Gray, Neville Brown, John Raymond, Philip Irwin, Brian Slater, Peter Hobbs, Des Cartwright, Allan Hogan, Mick Grixti, Des Dowling, all from New South Wales. Gordon Dosser, Graeme Hyde, David Judd, Peitre Vroegrop, Colin Cleak, Doug Ikin, Bill Shwarzenberg, John McCrory, all from Victoria. Reg Collyer, Max Niedorfer, Herb Forrest, Dave Farncombe, John Lewitzka, John Stracken, Max Peek, all from South Australia. Doug Anderson, Hank Jonker and Bob Phillpot from Western Australia.

A special acclamation is offered to Peter Brown for his unprecedented success with the Orange-bellied Parrakeet Recovery Programme as well as our thanks for the information provided.

To those who supplied us with photographs, many thanks. The list of photographers provides personal acknowledgement.

We are especially indebted to Barry Hutchins who kindly supplied us with much information and accepted our invitation to write the Foreword for this book.

A kind thank you to Rhonda Glover for the hours spent reading our handwriting and typing the manuscript.

Our long suffering wives Marilyn Gill and Jill Sindel must be thanked for the corrections and constructive criticism made during the tiring hours of proof reading as well as for the understanding shown to a couple of would-be authors.

We also applaud the journals of the avicultural societies in Australia, particularly *Australian Aviculture* and *Birdkeeping in Australia* as well as the journals of the Royal Australian Ornithological Union and the Bird Observers' Club of Australia, which all provide so much information for bird people of all persuasions.

Again we thank Ivor Beatty and his staff for their expertise and advice in helping us to print this book.

TABLE OF CONTENTS

LIST OF PLATES

Photographers — Reg Collyer: 3C, 19E; James Gill: 6A–D; Barry Hutchins: 11B, 13C; Dave Watts: 14, 15; All other photographs by Stan Sindel.

FOREWORD

Neophemas are a small group of the Avian treasures of the world, yet they are a source of fascination and affection for thousands of people. Their beauty and character set a unique example of nature and Australia has been blessed as the country of their natural environment.

For many decades conjecture has arisen concerning certain genera within the Neophema group. The authors with their findings and the results from research biologists have produced sufficient evidence to warrant changes. The taxonomic treatment is explained in the Introduction.

With the rapid progress of bird mutations in captivity during the past two decades in Australia it was inevitable that a publication on this selected group of birds would be published. The authors of this book have a wealth of experience in their own right and it is fortunate for Aviculture that Stan and James have combined to produce a publication not only on Neophema mutations but the coverage of all seven species in their natural entirety.

Stan Sindel has kept and bred birds for the past 49 years and is one of Australia's foremost aviculturists. He has qualities essential for a successful bird breeder and has always set high standards for bird husbandry. His eagerness to evolve additional ideas for improving avicultural techniques is once again evident in this book.

James Gill is a Veterinary Science graduate from the University of Sydney, 1976, and gained a Masters degree from the University of Glasgow in 1980. As a dedicated aviculturist for the past 29 years he has incorporated avian medicine, surgery and surgical sexing of birds into his Sydney practice.

Examples of amazing changes to the natural forms are displayed in descriptive detail of six of the seven species. These alterations from the originals place them in the category of mutations. All mutations known to the authors are listed and described where possible in excellent detail. This information will prove invaluable, as in time many of the mutations we are witness to today will be lost.

The explicit detail given to personal breeding results and careful observations once again gives merit to written notes rather than trying to rely on one's memory. To portray such information throughout this book for all to read must give the authors a sense of satisfaction. Perhaps such appreciation could encourage and stimulate more people to become better observers both in the field and with their birds in captivity.

An observation by Victorian Bill Schwarzenberg of the precocious behaviour of young Rock Parrakeets and the added notes by the authors on similar findings are some of the many enlightening sections in the book.

This book will join other worthwhile publications on field and aviary studies and bridge the gap between scientists and dedicated private collectors.

Adelaide, SA, 1992   Barry R. Hutchins
                                      Life member
                                      The Avicultural Society of
                                      South Australia Inc.

INTRODUCTION

Since the first exportations of Australian grass parrakeets (parrakeets have long tapered tails, whereas parrots have short square tails) during the nineteenth century, Neophemas have been sought after by birdkeepers throughout the world.

These elegant, diminutive birds with their soft, sweet voices are the ideal aviary psittacine, regardless of their keeper's situation. Even the delightful Budgerigar is noisy in comparison. Neophemas are so quiet and inoffensive that the domesticated species may even be kept and bred by high density housing dwellers without fear of disturbing their neighbours. They are everybody's bird.

This group has something for every aviculturist whatever their tastes; there are the brilliantly coloured Turquoise and Scarlet-chested Parrakeets, then in contrast the subtle, subdued colours of the Bourke's Parrakeet, and never has a bird been more appropriately named with its elegance of form and beauty than the Elegant Parrakeet.

For those who like more challenging species, the Blue-winged Parrakeet to a point, and the Rock Parrakeet are ideal and packed with interest. Then there is the rare and endangered Orange-bellied Parrakeet, *Neophema chrysogaster,* whose future is uncertain. A captive breeding programme is now in progress but it is quite unlikely this species will ever be kept by private aviculturists in Australia even though the enlistment of our expertise may be the only way to increase numbers quickly.

The fact that most species are prolific has assisted the establishment of numerous and varied colour mutations in many species, thus adding to the avicultural interest in this group. Has there ever been a more beautiful colour mutation than the male Dilute Yellow Turquoise Parrakeet?

At first glance the seven species usually included in the Neophema genus appear to be closely allied, and indeed some features such as similarity of shape, size and feeding habits as well as the six central feathers of the

11

tapered tail being of similar length, suggest this. Yet further investigation, particularly at an avicultural level, indicate three distinct groups extending to two genera within this genus.

Only two species conform closely to the typical Neophema genus in which sexual dimorphism is prominent, the Turquoise Parrakeet, *Neophema pulchella* and the Scarlet-chested Parrakeet, *Neophema splendida*. The almost total lack of records of fertile hybrids from between these species suggests even they are not closely allied.

Four species form the blue frontal band group in which sexual dimorphism is barely apparent or non-existent.

This group was previously the genus *Neonanodes* and are currently, in order of the extent of specialisation, the Elegant Parrakeet, *Neophema elegans;* the Blue-winged Parrakeet, *Neophema chrysostoma;* the Orange-bellied Parrakeet, *Neophema chrysogaster;* and the most specialised of all, the Rock Parrakeet, *Neophema petrophila*.

The apparent inability of any of these species to produce fertile hybrids with either of the previous group (i.e., the typical Neophemas) suggests only distant relationship with the typical Neophemas, which some authorities believe warrants the reintroduction of the *Neonanodes* genus.

Fertile hybrids within the Neonanodes group have been recorded suggesting close relationship within the group.

The sole member of the third group and what has now proven to be a distinct genus is the Bourke's Parrakeet, *Neophema bourkii,* previously *Neopsephotus bourkii*.

Research instigated by the C.S.I.R.O. under the title "Biochemical Systematics within the Australo-Papuan Parrots, Lorikeets and Cockatoos", in layman's terms examines blood protein variations within the Australian Psittacine genera and species using an electrophoresis technique.

This ongoing research is carried out by: L. Christidis, Curator, Department of Ornithology, Division of Natural History, Museum of Victoria; R. Schodde, Australian National Wildlife Collection, Division of Wildlife and Ecology, C.S.I.R.O.; D. D. Shaw and S. F. Maynes, Molecular and Population Genetics Group, Research School of Biological Sciences, Australian National University.

Les Christidis explained to us that the results of this research places the Bourke's Parrakeet in a monotypic genus which is as remote from all other Australian Psittacines as the Budgerigar, *Melopsitticus undulatus,* and the Cockatiel, *Nymphicus hollandicus.*

On this evidence we are reintroducing the monotypic genus *Neopsephotus* to accommodate the Bourke's Parrakeet as *Neopsephotus bourkii,* in the genus created for it by Gregory Mathews in 1913 after recognising this species was distinct from the *Neophema* genus.

Summarising our taxonomic treatment of this group: Typical Neophemas — Turquoise Parrakeet, *Neophema pulchella;* Scarlet-chested Parrakeet, *Neophema splendida.* The Blue Frontal Band Group (previously *Neonanodes*) — Elegant Parrakeet, *Neophema elegans;* Blue-winged Parrakeet, *Neophema chrysostoma;* Orange-bellied Parrakeet, *Neophema chrysogaster;* Rock Parrakeet, *Neophema petrophila.* Neopsephotus Genus — Bourke's Parrakeet, *Neopsephotus bourkii.*

With the exception of the Orange-bellied Parrakeet, the members of this group, although not common, appear to be holding their own in the wild when factors such as remoteness of habitat and specialisation are considered.

At an avicultural level in Australia this group, again with the exception of the Orange-bellied Parrakeet, is successful, secure and self supporting within the avicultural industry. Yet still the occassional report of trapping of these species persists.

The trapping of any native bird for illegal trade can only help to tarnish the image of aviculture — an image the majority of us are fighting so hard to elevate. Trapping of members of this prolific group is so unnecessary and only serves to nullify the rewards of those aviculturists who strive for breeding success.

The difficulties created by illegal trapping for aviculture and aviculturists are relatively minor when compared to the damage caused by the trapping to localised populations.

Finally we continue with the aim of our specialised books to provide new and experienced aviculturists with a complete understanding of the birds they keep, using data relating to the past, the present and perhaps the future. All information is presented and new techniques explained simply and in language well within the range of all our comprehensions. Avicultural or ornithological words or terminologies which may not be understood by newcomers to aviculture are defined on their initial use.

13

HOUSING

Most species of Neophema have been housed and bred in every conceivable type of aviary and cage, so obviously housing for most members of this group is not as critical as it is for other genera of Australian Psittacines.

We have personally bred Bourke's, Turquoisines, Elegants, Blue-wings and Scarlet-chesteds in large planted aviaries ranging from 4 m (13 ft) to 12 m (40 ft) in length. Conversely we have bred Bourke's, Turquoisines and Scarlet-chesteds in small cages or cabinets approximately 1 m (3 ft 3 in.) long, 50 cm (1 ft 8 in.) wide and 50 cm (1 ft 8 in.) high.

When contemplating aviaries for the Neophema genus, where the majority of species are totally domesticated, quite prolific and tolerant of a wide variety of housing, our consideration must be extended to simplify management and eliminate unnecessary losses. Although Neophemas are relatively steady aviary birds, individuals will occasionally take fright at some unseen danger and crash into the wire mesh of the aviary in a frenzied effort to escape. This practice often results in death from a fractured skull or broken neck.

Over 30 years ago Stan witnessed such an event while enjoying the sight of his first and at the time, only pair of Turquoisines feeding from seeding grasses on the floor of the open flight of their aviary. Suddenly the male took fright and flew straight up into the wire netting on the top of the flight, then fell to the floor dead. This is one of numerous hard lessons Stan has learnt from his birds over the years and one which can be overcome with safe sensible housing.

To combat such unpredictable behaviour as well as the common practice of roosting at night in open flights regardless of how severe weather conditions may be, to reduce worm infestation from wild bird droppings and to prevent predator attacks, we believe all Neophema aviaries should be fully roofed.

Unlike most parrot groups, the domesticated species of Neophema may be housed in breeding cabinets for long periods without adverse effects, providing a suitable non-fattening diet is used and at least two months of flying in a larger aviary is given annually.

Although this group of small grass parrakeets who feed on or predominately near the ground, seem an unlikely genus to house in suspended wire aviaries or even cages, numerous breeding successes have been recorded from a number of species housed in these conditions. From the collation of all available breeding data there emerges four distinct types of successful housing for Neophemas; larger type planted aviaries, smaller sheltered aviaries, suspended wire aviaries and breeding cabinets.

The Larger Type Planted Aviary

These aviaries are far too varied to suggest a particular shape or size but the following points should be considered.

(a) The aviary should be totally enclosed on the southern and western sides to provide maximum protection from prevailing winds.

(b) Shelter areas should have concrete floors to simplify cleaning.

(c) The aviary should be provided with as much protection as practicable on the sides with solid materials such as iron or fibreglass sheeting.

(d) Flight patterns within large planted aviaries should be broken up with shrubs to slow down the Neophemas flight speed, thus reducing self inflicted injuries.

(e) Aviaries should be totally vermin proofed by providing foundations deep enough to prevent rats and mice from tunnelling into the aviary. Flat metal sheeting should extend from ground level to at least 60 cm (2 ft) high to exclude rodents and at least twice this height in snake prone areas.

(f) Construction materials for larger aviaries housing only Neophemas or together with finches and softbills is not critical as chewing damage inflicted by these grass parrakeets is only minimal.

The Smaller Sheltered Aviary

This form of housing must be considered ideal for most species of Neophema (see Figure 1).

Preferably this type of aviary should be totally enclosed except at the front, fully roofed with solid walls and concrete floors.

16

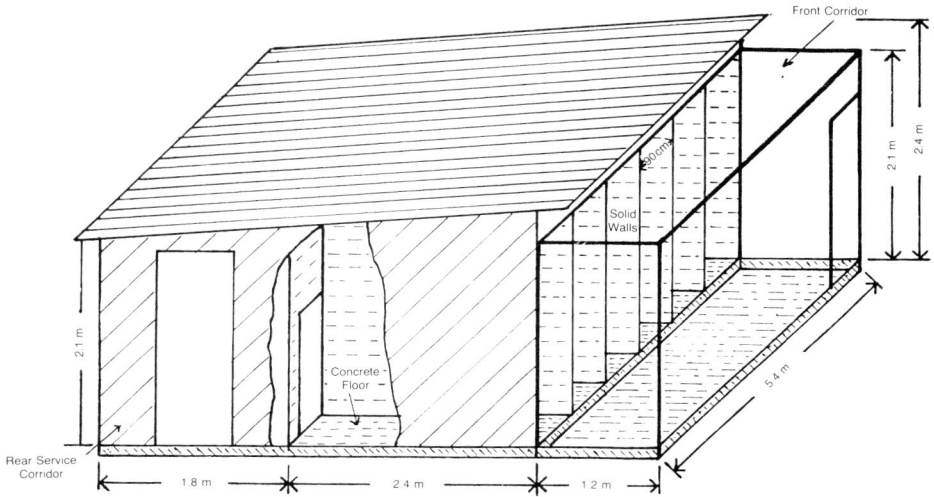

Figure 1. Fully roofed aviary block with rear service corridor and front corridor or planted area for protection of aviary fronts.

If such aviaries are intended to house only Neophemas the types of building materials used are not as critical as those used for the larger parrots. Even soft woods are suitable for members of this genus.

The following features should be incorporated in the design of these small sheltered Neophema aviaries.

(a) Aspect

In the southern hemisphere this type of aviary should always face north to maximise the amount of direct sunlight received.

(b) Framework

Currently the trend in Australia is moving towards the construction of all aviary frames with galvanised metal tubing of either round or square section.

Welding the joints of the tube framing is the cheapest and most efficient method of construction, although some degree of skill is necessary.

Various tee and elbow joints are available for both square and round tubing and although expensive, require far less skill to assemble.

Square tubing may be joined efficiently by cutting and folding the end of one section so as to receive another and then securing the lapped joint with pop rivets or tech screws.

For aviaries that are to be used to house only Neophemas, finches and softbills easy to work softwoods such as pine can be used for framing construction.

Australian hardwood framing will withstand the chewing of larger parrot species such as Psephotus, and even the Rosella family (Platycercus) will inflict only minor damage.

(c) Walls

The use of brick walls eliminates the need for wall framing and obviously provides very substantial aviaries, but when the costs of brick construction are added to the costs of the necessary supportive concrete foundations, brick walls are not viable for small parrot breeding aviaries.

Metal wall sheeting, either painted or colour bonded, provides an easy to clean, non porous surface that is easily fixed to any type of framework.

Fibro cement provides a cheap and effective wall sheeting, although the porous surface is more difficult to clean unless painted with a gloss paint. This type of sheeting will withstand the chewing of much larger parrots than Neophemas but is subject to damage or break-ins when used on external walls. Fixing on all types of framework is relatively simple.

Plywood sheeting is easy to fix to all types of framework but its porous surface makes it harder to clean unless painted with gloss paint. Neophemas can inflict minimal damage if allowed access to the edge of the sheet.

(d) Roofing

Metal roofing is a durable material which, unlike other roofing materials, when fixed correctly is almost totally immune to storm damage. In extreme climates it is advisable to install insulation under the metal roofing to moderate temperature within the aviaries. Heat is reflected by the bright surface of galvanised, zincalum or aluminium sheeting or by roof surfaces painted white, thus reducing summer temperatures within the aviaries.

Stan has found that metal roofing fixed on top of flat fibro sheeting provides added insulation and thus reduces the temperature inside his aviaries during the hot summer months. Conversely winter temperatures are higher within his aviaries. The fibro cement sheeting also provides an easy to paint ceiling under the metal roofing.

All other types of roofing materials available in Australia are unsuitable for aviary construction for various reasons such as vulnerability to storm damage, leaking, costs, etc.

(e) Wire Mesh

Almost any form of mesh of suitable size is capable of containing all species of Neophemas, which leads to speculation that nylon or plastic meshes could reduce injuries caused by birds, particularly fledglings, flying into the mesh front of their aviary. These nylon meshes have been used in some public softbill aviaries but we have had recent reports of rats chewing through the mesh and birds escaping.

Mesh sizes of 12 mm (½ in.) square or 25 mm (1 in.) by 12 mm (½ in.) are ideal, regardless of the material.

Galvanised 12 mm (½ in.) "bird" wire is commonly used for small parrot aviaries but we feel that 25 mm (1 in.) by 12 mm (½ in.) galvanised weldmesh of 17 Gauge is the most suitable mesh currently in use.

Methods used to fix mesh must avoid any possibilities of accidental injuries. The mesh must be cut carefully so there are no jagged edges. If wire ties are used they must not have long ends which may catch rings or legs. Pop Rivets with large washers are an exceptionally safe method of attaching wire to a metal frame.

(f) Partitions

Most Neophema aviaries are built in groups or blocks, and when this is the case the partitions between the individual aviaries may be of a solid material as recommended for walls or wire mesh. Solid partitions eliminate interference from neighbouring pairs and provides a draught free secluded environment inducive to successful breeding.

Wire divisions between aviaries require double wire mesh fixed at least 25 mm (1 in.) apart, to avoid injuries being inflicted through fighting during the breeding season. Although most Neophemas are relatively peaceful some individuals can be quite aggressive at this time.

We feel the advantages of solid partitions far exceeds those of the wire partitions although the use of wire partitions in areas of extreme summer temperatures has the advantage of assisting air circulation. Similar conditions can be achieved in solid partitioned aviaries by installing ventilation in the rear of the aviaries.

Floors

The most commonly encountered floors in small parrot breeding aviaries are natural earth, sand filled, pebble and concrete.

Natural earth floors provide a pleasing appearance in larger planted aviaries, but in smaller parrot aviaries the earth soon becomes polluted and unhygienic. Birds held on this type of floor must be wormed at least every three months and are vulnerable to infections from many types of bacteria which thrive in damp earth. This type of floor is also subject to rodent infestation.

Sand filled floors provide a pleasing appearance and are reasonably easy to clean by skimming droppings off the surface. We believe it is necessary to replace the top 10 cm (4 in.) of sand every 12 months, for if the sand is allowed to remain in an aviary over long periods of time it will silt up with dust, fine excrement particles, seed husks etc., and becomes almost as polluted as earth floors. Birds held on sand floors also require regular worming. Drainage of sand filled floors is essential, either in the form of sub surface agricultural drainage (see Figure 2) or by elevating the floor level to at least 10 cm (4 in.) above the surrounding ground level and providing suitable seepage points.

Pebble floors also present a pleasing appearance and if provided with an efficient sub surface drainage system (see Figure 2) may be cleaned easily by high pressure hosing. If pebble floors are not provided with a drainage system the surface is difficult to clean and silting up is inevitable. Even adequately drained pebble floors are subject to silting up after several years of use, necessitating renewal of the pebbles. Regular worming of birds housed on pebble floors is essential.

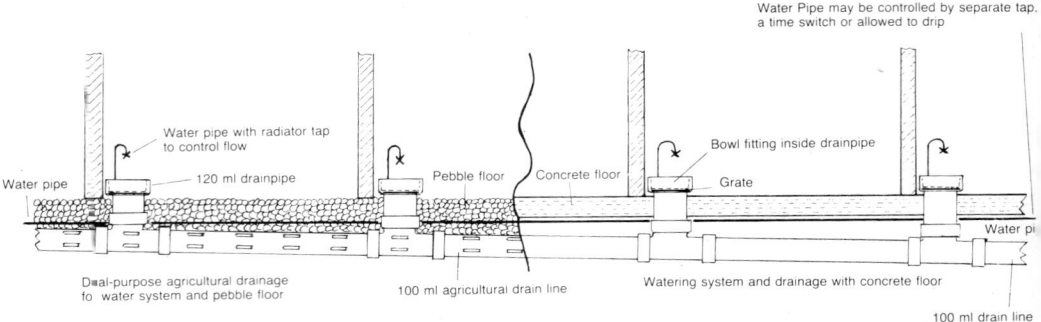

Figure 2. Watering system.

Concrete floors are the most functional, easy to clean and maintain type of aviary floor, although perhaps of a less pleasing appearance. Cleaning of this floor is simplified by a liberal sprinkling of dry sand which prevents droppings and stale food stuffs from adhering to the surface. The use of a broom and shovel is all that is required to remove the bulk of the waste. Floors with an adequate drainage may be cleaned by using a high pressure water hose. Concrete floors also eliminate vermin such as mice and rats as well as helping to control worm infestation.

Walkways and Service Corridors

All aviary complexes should ideally be provided with front access walkways or fully roofed rear service corridors or both (see Figure 1). Walkways or service corridors situated at the front of aviary complexes eliminate escapes, predator attacks and general disruption, as well as facilitating feeding and cleaning.

Fully roofed service corridors at the rear of an aviary complex provides even easier feeding and cleaning facilities under all weather conditions. This type of service area when designed with sufficient width allows the use of external feed hoppers, watering devices and nesting logs or boxes, all of which provide easy access, quick servicing and the minimum disturbance of breeding pairs. The only disadvantage we can see with the rear service corridors is the vulnerability of the front of the aviaries to predator attack and other disturbances which are eliminated by front service areas. However, this problem can be overcome by the construction of a front access corridor as well as the rear corridor, or preferably, even if just for the aesthetic value, a planted aviary at the front of the breeding aviary complex (see Figure 1).

Front access service corridors should be a minimum width of 1.2 m (4 ft) to provide comfortable working conditions whereas rear access service corridors should be a minimum of 1.8 m (6 ft) wide.

Advisable Minimum Aviary Size

The ideal dimensions of an aviary to house a pair of Neophemas for breeding purposes is 2 m (6 ft 6 in.) to 3 m (9 ft 9 in.) long, 90 cm (3 ft) to 1.2 m (4 ft) wide and 2 m (6 ft 6 in.) high with our own preference being for no more than 2.4 m (8 ft) in length.

Colony aviaries for those species that do well in colony situations such as Blue-wings and Rock Parrots, ideally should also be fully roofed and range from 3 m (9 ft 9 in) to 4 m (13 ft) in length, 1.2 m (4 ft) to 1.8 m (6 ft) wide and approxiately 2 m (6 ft 6 in.) high.

Feeding and Watering

Feeding and watering facilities must be located so as not to be polluted or contaminated by the birds droppings as well as being protected from rain, direct sunlight and vermin. These requirements may be achieved in numerous ways but basically there should be no possibility of birds perching above feed trays or water dishes. Both the water and feed containers must be under shelter and the food containers suspended in such a way as to prevent access to mice and rats.

Snake Proofing

Aviculturists who live in rural areas of Australia are liable to be troubled by snakes in their aviaries at some time or another. Neophemas are particularly vulnerable because of their small size which means that not only are the eggs and young at risk of snake predation in the nest, but the adult birds may also fall prey to even a medium sized snake.

Being city dwellers we have no first hand experience of snake predation but all those who have experienced the problem agree that snakes are incredibly difficult to exclude from an aviary. Snakes can squeeze through the tiniest crack, come in through mouse holes, get through 12 mm (½ in.) mesh and scale vertical walls where the slightest protrusion may aid their ascent.

Smooth surfaced metal sheeting fitted around the lower walls of an aviary to a height of 1.2 m (4 ft) helps to exclude them.

Electric cattle fences are now being used successfully as a snake deterrent. The electrified wire is installed close enough to the side of the aviary to make it impossible for the snake to pass between it and the wall without making contact. Likewise if the snake passes outside the wire it will also make contact.

Suspended Aviaries

A suspended aviary is one constructed of wire mesh and suspended at a suitable height above ground level to facilitate cleaning underneath.

This type of housing is used extensively overseas for numerous parrot species. In Australia it has been used successfully with native and exotic species and we use this system almost exclusively for our lorikeet housing.

Although the concept of housing Neophemas, who are ground frequenting parrakeets, in suspended aviaries is difficult for some to accept, it is proving successful in many collections with a number of species.

Suspended aviaries are best located in vermin and bird proof enclosures, inside large aviaries or with doors opening into secured bird proofed service areas to facilitate feeding etc. and to eliminate the possibility of escape.

When this type of housing is used for Neophemas, the aviaries should be situated in a well protected draught free and totally roofed area which faces north and allows direct sunlight to reach at least the front section.

Alternatively if the aviaries are suspended in an open area they must be protected from prevailing winds and provided with individual shelter, preferably at both ends or possibly fully roofed (see Figure 3).

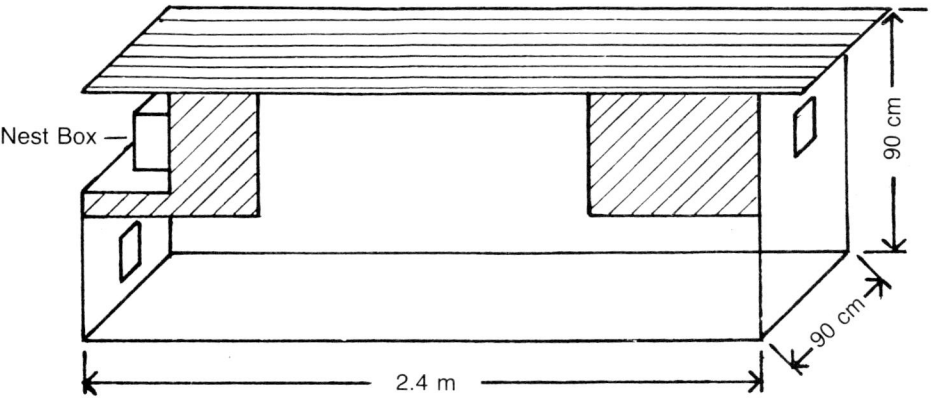

Figure 3. Suspended aviary with protecting shelter and full roof.

We are not convinced that this type of housing is ideal for Neophemas but we have records of succesful breedings of Bourke's, Turquoisines, Scarlets, Elegants, Blue-wings and Rocks under these conditions.

The advantages and disadvantages of this type of housing are as follows.

Advantages

1. Provides an easy to clean and hygienic environment.
2. Eliminates the birds access to their droppings and to stale food items, all of which fall through the bottom of the aviary.
3. Facilitates servicing with the minimum disturbance to the occupants.
4. External nesting sites simplifies their cleaning and inspection without disturbance to the birds.

5. Worming is unnecessary when the birds are housed in suspended aviaries, providing they are free of worms to start with.

6. The cost of erecting suspended aviaries is far less than for conventional aviaries.

7. More birds can be housed in a limited area.

8. Some birds feel more secure because the aviary cannot be entered by the keeper.

Disadvantages

1. Provides little protection from draughts or winds, particularly when situated outside enclosed areas.

2. Birds such as Neophemas which will not roost in their nesting facilities are adversely affected by cold in extreme climates. The microclimate in the south western suburbs of Sydney where we both live makes suspended aviaries impractical for use with Neophemas, for this reason.

3. Wire floored aviaries provide an unnatural environment for ground frequenting birds such as Neophemas. This can be compensated for by providing a tray of turf which is changed frequently.

4. Efficient watering systems are difficult to install in suspended aviaries.

Battery systems of suspended aviaries which have single wire partitions may require the juggling of some pairs in order to obtain compatible neighbours. Should fighting, maiming or excessive interest in a neighbour disrupt breeding, double wire or solid partitions may be necessary. Single aviaries suspended several centimetres apart provides a viable alternative.

Suspended aviary dimensions may range from 1 m (3 ft 3 in.) to 3 m (9 ft 9 in.) long, 60 cm (2 ft) to 90 cm (3 ft) wide and 60 cm (2 ft) to 90 cm (3 ft) high.

Cabinets

Firstly we must stress that breeding cabinets are not a long-term housing solution, but as the name suggests, are for the duration of the breeding season only.

During 1984, Stan was a member of a group of Australian aviculturists who toured Europe and England. Whilst in England Stan and some of the group visited the small backyard collection of an aviculturist who had been retired for some years. He devoted most of his time to his birds and specialised in Neophemas.

All his breeding was carried out in cabinets which measured approximately 1.2 m (4 ft) long, 60 cm (2 ft) high and 60 cm (2 ft) deep.

The species this master aviculturist bred in these cabinets were Turquoisines (which were his speciality species, that he had bred for 30 years and were the best that Stan had ever seen), Scarlet-chested, Elegants, Blue-wings and Bourke's.

The cabinets were situated in partially open fronted, bird proofed buildings which were fitted with automatic lighting designed to provide 14 hours of light per day, all year round. He maintained that all Neophemas if given 14 hours light per day will breed, regardless of the season and even in an English winter. In other words he believed that Neophemas were stimulated into breeding condition by increased daylight hours regardless of temperature. Hence he used a winter and summer breeding team, with each being rested in large aviaries during the six months non breeding season.

This man produced more quality Neophemas in his small collection each year, than any dozen Australian collections of similar size.

We have documented a number of extremely sucessful breeding pairs of Neophemas housed in cabinets in Australia, these are dealt with in the species chapters.

Cabinets should be situated in open, wire fronted shelters which face north and allow direct sunlight to reach them, or if situated outdoors should have a weather-proof roof. Alternatively they can be housed in birdrooms with artificial light provided.

Sliding sheet metal trays on the floor of the cabinets facilitates cleaning and when a wire mesh false floor is fitted approximately 25 mm (1 in.) above the tray, access to droppings and stale food is eliminated.

Breeding cabinets ideally should be constructed of a lightweight non porous material such as fibreglass and slotted into shelves or racks. Such construction allows the cabinets to be removed bodily for cleaning and disinfection. Most cabinets are constructed as fixtures and sheeted with porous materials such as fibro cement, plywood or particle board. These cabinets are impossible to remove individually and hence more difficult to clean although if coated with a non toxic high gloss paint the task is made easier. When designing breeding cabinets thought must be given to cleaning, feeding, watering and external nesting facilities so as to minimise disruption to the breeding pairs.

Neophemas are nervous birds by nature and the cabinets should be no narrower than 60 cm (2 ft) so as to allow the birds to feel comfortable. Outside disturbances should be kept to a minimum.

DIETS

The task of providing a nutritious, well balanced diet for the members of the Neophema genus is less daunting than for many other parrot groups.

In the wild, Neophemas have been reported to feed on the seeds of various grasses (both native and introduced), herbaceous plants, shrubs and trees, vegetable matter, flowers, blossom, berries and small fruits as well as a few insects and their larvae.

Numerous diets have been developed to suit the five domesticated species of this genus but particular attention must be paid to the diet of the specialised Rock Parrakeets and the Orange-bellied Parrakeets in captivity.

We have found all Neophemas can be subject to obesity to varying degrees in Australian aviaries, particularly when housed in a small breeding aviary. Rock Parrakeets can become particularly obese and lethargic in aviaries without a suitably controlled diet and Stan experienced similar problems with the Orange-bellied Parrakeets held in his aviaries over 20 years ago.

To combat obesity we suggest a non-oily, low protein, basic dry seed diet which is best provided by any of the millet seeds. After a number of years of experimentation with various types of millets, we found the most suitable and preferred to be French White Millet. Neophemas all preferred this millet above other varieties, hence wastage was less and nutrition at least equal to that of other millets.

We believe that dry seed should only be supplementary to the daily supply of greenfood, sprouted seeds, vegetables, fruit, etc.

The feeding of fattening or oily seeds such as sunflower seed, canary seed, oats etc., should be on a very limited basis or preferably totally avoided except when young are being reared.

Fat birds cannot be considered healthy birds or potentially good breeders as obesity adversely effects fertility and egg laying as well as promoting heart disease.

Hence the ideal daily diet should consist of suitable greenfood, palatable vegetables, fruit and sprouted seeds all of which are supplemented with a non-oily, low protein dry seed such as French White Millet.

Greenfood

Fresh greenfood is a highly nutritious, non-fattening food source.

Being a city dweller, Stan relies heavily on purchased greenfood such as silverbeet, endives, chicory etc. All of these are thoroughly washed before use to ensure the removal of insecticides. Stan will only feed natural greenfood such as seeding grasses etc., that are picked in the confines of his own property or in other known safe areas as the use of herbicides by local councils in streets, parks and other public places makes collecting greenfood in these areas dangerous.

On the other hand, Jim, who resides on acres in a rural suburb of Sydney, has the advantage of being able to harvest large quantities of natural greenfood on his own property.

Many aviculturists cultivate their own greenfood in small plots of domesticated varieties. The cultivation of seeding grasses such as winter grass, summer grass, water millet, milk thistle, chick weed, dandelion, pig weed etc., which are all superb greenfoods, is advisable.

Sown grains such as all the millets, milo, oats, barley, canary seed, etc., all provide good seed heads, which when fed in the milky stage are a palatable and highly nutritious food source.

Vegetables

Fresh vegetables are another highly nutritious food source which should be exploited, although many Neophemas refuse to eat most varieties. Cauliflower, broccoli, green peas, fresh beans and carrots are all worth trying. Fresh peas are by far the most favoured with all species and are particularly relished when young are being reared.

Frozen peas are also taken by most birds but with less enthusiasm, obviously doubting the nutrition retained in frozen vegetables — just as we do.

Fruit

Neophemas are not particularly fond of fruit although some species, particularly Scarlet-chested and Rock Parrakeets are partial to a slice of apple. Any fruit that an individual or species show even a slight interest in should be persevered with.

Sprouted Seed

Sprouted seed provides a constantly available, highly nutritious food source which contains many of the essential vitamins such as niacin and riboflavin.

Most varieties of seed may be sprouted safely by soaking for 24 hours in water to which has been added an anti bacterial, anti fungal agent such as Sodium hypochlorite. We use the formulations which are used to sanitise babies bottles. Some aviculturists use household bleach for this purpose but Jim feels that the lack of quality control imposed on these bleaches renders their use undesirable. Sunflower seed will sprout more successfully after only three hours of soaking. Longer periods increase the chances of contamination particularly in hot weather.

The soaked seed is then placed in a perforated plastic ice-cream container or colander, thoroughly washed and drained, then left to sprout in the container. This process may take from 24 hours to four days according to the type of seed and the temperature they are sprouted at. Sprouting usually takes one or two days longer during the winter than the summer, but may be hastened by placing a lid or sheet of glass over the container of seed.

There is no advantage in allowing sprouted seed to grow long shoots as the maximum nutritional value is reached when the shoot breaks out of the seed case. Also most birds are only interested in eating the kernel, not the shoot. Sprouted millets may be fed on an unlimited basis but sprouted oily and fattening seeds such as sunflower seed, canary seed or oats should be limited to a dozen or so seeds per pair per day during the non breeding season. This is to avoid them becoming obese. These sprouted fattening seeds may be gradually increased a few weeks prior to the breeding season to help stimulate breeding and then fed on an unlimited basis when young are being reared.

We have not been able to find any original research on the nutritional changes that take place during the germinating process. However, the birds relish sprouted seed and our observations indicate an improved performance when fed sprouted seed.

Storage of Seed

Seed should be stored in a dry environment in specially constructed metal bins, metal or plastic drums or similar containers which have close fitting, air tight and vermin proof lids.

Storage of Perishable Foods

The storage life of greenfood, fruit, vegetables and sprouted seed is extended considerably when kept under refrigeration. During periods of hot weather the rapid growth of sprouted seed is inhibited and souring prevented by refrigeration.

Rearing Foods

As soon as babies hatch, breeding pairs must be supplied with additional nutritious foodstuffs to enable them to rear healthy, robust youngsters.

Greenfoods, vegetables, seeding grasses, half ripened cultivated seed heads and sprouted seeds are all important and desirable rearing foods. Milk arrowroot biscuits, madiera plain cake, egg and biscuit canary rearing food moistened to a crumbly consistency and moistened chicken crumbles also provide additional and easily obtainable, nutritious rearing foods.

Pairs who are rearing young should be monitored to ascertain exactly what they are eating and hence what they are feeding to their young. If there is doubt as to the quality and quantity of food being consumed, other suggested rearing foods should be provided.

Stan has found all his Neophemas will rear successfully when supplied with sprouted sunflower seed and sprouted French White Millet, Silver beet, fresh green peas, milk arrowroot biscuits, a little plain cake, cuttlefish bone and whatever seeding grasses are available (usually only limited amounts are available). Dry sunflower seed and French White Millet is also available throughout the rearing period.

Jim uses a similar formula but with a greater emphasis placed on natural seeding grasses which are supplied fresh daily in large quantities whenever available.

Calcium

It is important that Neophemas have adequate calcium at their disposal, particularly prior to egg laying and while young are being reared.

Cuttlefish bone is the usual and most convenient way of supplying calcium to this group as it is relished by all species. Calcium grit and mineral blocks are also useful for this purpose.

Charcoal

Charcoal is appreciated by Neophemas particularly when young are being reared.

Commercial Parrot Pellets

Pelleted parrot feed is becoming available and will become increasingly acceptable to aviculturists. They aim to provide the birds nutritional requirements in a single convenient formulation.

The basic drawback with the principle is that we rarely know a particular parrot species' nutritional requirements. If the birds have only the pellet to choose from they have no chance of balancing the diet for themselves. Pheasants and turkeys do not thrive on the pellets of the more common galliform, the fowl. It is also asking a lot to accept that a Bourke's Parrakeet will do well on the same pellet that a Rock Parrakeet will do well on.

Intelligent birds like parrots may well become bored with a pelleted diet.

Pelleted feeds certainly have a future but there is a long way to go in our understanding of the parrot's nutritional requirements before they can be accepted as a blanket answer.

Jim is using one formulation as a supplement to his parrots' diet.

During a visit to New Zealand Stan met a highly competent aviculturist who placed his large and varied collection totally on a pelleted diet. This is the way in which the pellets are recommended to be used — not supplemented by fruit vegetables or sprouted seed.

The following breeding season was the worst he had ever experienced with most pairs not attempting to breed and the remainder having poor results. The pairs' production was down in excess of 90%.

He reverted to his original feeding programme and the ensuing breeding season exceeded normal production.

Hand Rearing Diet

There are numerous hand rearing diets in use throughout the bird world. Most aviculturists have a hand rearing mixture that they swear by, that is until something goes wrong. Then they blame their diet and abandon it for some newly found super rearing diet.

In fact, most failures in hand rearing are not related to the failure of a proven diet but to other factors such as bacterial infections, numerous other diseases, poor brooding facilities and so on.

Our requirements for a hand rearing diet are:
1. It is simple to prepare.
2. It should have no ingredients that will deteriorate or promote bacterial or fungal growth when stored.

31

3. It is a dry mixture requiring no cooling or storing under refrigeration.
4. It should provide all the nutrition required to rear healthy chicks.
5. It should have a 15–20 per cent protein content which appears to be the optimum range for rearing parrot chicks.

Many hand rearing diets either meet or go close to meeting these requirements, with those diets requiring cooking being adequate although time consuming and inconvenient for the user. We would suggest that if you have a diet that works for you, don't change it.

Hand Rearing Diet

1 part ground chicken starter crumbles (Approximately 20 per cent protein).

1 part ground egg and biscuit canary rearing food.

1 part sunflower meal.

1 part Farex Baby cereal.

1 level teaspoon of multi vitamin and mineral powder such as Ultrameal or Ornithon should be added to each four cups of the mixture.

This dry mix is best stored in an airtight container and kept in a cool dry place.

To prepare for feeding, simply add hot water and allow to cool to a suitable temperature so as not to burn the chicks. A calcium supplement such as Calcium Sandoz or DCP340 powder should be added to one feed each day and a few drops of liquid vitamin supplement such as Penta Vite or Avi Drops to another.

Hand Rearing Diet For Day Old Chicks

During the last three years Stan has tested a variation of the hand rearing diet published by Hutchins and Lovell in their *Australian Parrots* for the rearing of day old chicks. It was found to be the most reliable and easily digested hand rearing diet he has used for day old chicks. The growth rates are slow but the addition of a greenfood puree increased this.

The preparation is as follows:

Blend together and put in a saucepan:

1 heaped tablespoon Wheatgerm

1 heaped tablespoon Heinz high protein baby cereal

1 small level teaspoon calcium powder (DCP340)

A pinch of salt

32

Place an egg yolk in a cup with a level teaspoon of honey and thoroughly stir in hot previously boiled water to fill the cup. Then add to the dry ingredients in the saucepan and stir into a wet porridge. Bring to the boil and simmer for four minutes while stirring. When cooked pass through a fine baby food sieve.

The resulting mixture may be stored in an airtight glass jar under refrigeration for a maximum of four days.

Greenfood Puree

To prepare take the peas from six pods of fresh green peas and approximately one hundred square centimetres of fresh silverbeet and blend into a cup of warm boiled water. The resulting liquid is passed through a fine baby food sieve and then stored in an airtight glass jar under refrigeration for up to four days.

See the management chapter for hand rearing techniques.

Commercial Hand Rearing Preparations

Of the few commercially prepared hand rearing diets currently available in Australia "Roudybusch Formula 3 Handfeeding Formula for birds" is the most prominent.

We have heard good reports with this diet even with day old chicks.

Stan's experience has been limited to incubator hatched Macaw chicks which did not do well on the diet, and poor weight gains forced him to revert to the above diet after three days. Subsequently incubator hatched Black Cockatoo chicks have done well on the Roudybusch formula up to ten days when they were weaned on to the home made diet.

When the opportunity arises we will test and compare both diets on less valuable chicks and submit the results in our next publication.

MANAGEMENT

Most species of Neophemas are now totally domesticated, and as such their requirements in captivity are well established. Nevertheless a strict management routine, sensible diet and clean, protected housing are essential for continued successful results with this group.

All captive livestock must be kept in clean conditions if they are to remain healthy. Weekly sweeping or scraping of floors, particularly under perches, is essential. Where floors are suitably drained, high pressure water hoses can be used to advantage.

The regular scrubbing of floors, walls, wire mesh, perches, food and water containers etc. with an iodine based disinfectant such as "Sani-Chick", will go a long way towards controlling disease outbreaks. Periodic liming of floor surfaces is also helpful in this area.

Perches should be positioned at the extremities of the aviary to provide maximum flying as well as being placed to eliminate the fouling of food and water receptacles. Either dowel sticks or natural branches may be used for perches — dowel sticks have the slight advantage of being easier to clean.

We have found an easily made perch holding bracket simplifies the fixing and replacing of perches in solid partitioned aviaries. A 2.5 cm (1 in.) wide "V" is cut in the centre of one side of a 10 cm (4 in.) length of 18 mm (3/4 in.) aluminium angle mould. The "V" cut out is then folded together to form a "V" shaped piece of angle mould, fixing holes are drilled near each end and the bracket fixed to the wall in the desired position with screws or pop rivets (see Figure 4).

Water Containers and Facilities

Neophemas seldom bathe, an adequate water dish approximately 15 cm (6 in.) to 20 cm (8 in.) in diameter and 3.7 cm (1.5 in.) to 5 cm (2 in.) deep should be supplied for a pair or small colony. Larger dishes are more liable to fouling and can be dangerous for newly fledged young.

Figure 4. Perch holder.

Water containers should have a smooth, non-porous, glazed surface to facilitate cleaning and reduce the growth of algae. We find the large china soup plates which are often available in supermarkets at a reasonable price are ideal for this sized parrakeet.

When any of the automatic watering systems are used it is essential to install a suitable drainage system to eliminate continually wet floor areas (see Figure 2). The water supply may be controlled manually at each bowl or by one control tap for several aviaries.

Fully automatic watering systems utilising an electronic valve and time switch on the water supply to turn on the water once or twice daily are time and labour saving devices which ensure regular filling of the water dishes.

Feeding Facilities

We prefer a two or three seed compartment hopper fitted with a tray which serves the dual purpose of catching the seed husks as well as providing a place to feed the supplementary foods (see Figure 5). This hopper and tray is constructed of sheet metal with a glass or perspex front to the seed compartments to facilitate the monitoring of seed levels, and provides an easy to clean, labour saving device which works well for Neophemas.

If seed is provided in dishes, they should be glazed for easy cleaning and placed in suspended metal trays, such as inverted metal garbage bin lids, which retain the husk and scattered seeds as well as catering for the supplementary foods.

All feeding facilities must be located in a totally dry, and weather protected position which also excludes the access of vermin such as mice.

Weekly emptying and cleaning of trays and feeding utensils is essential, as stale food and dusty, mouldy or fouled seed can promote fungal and bacterial infections.

36

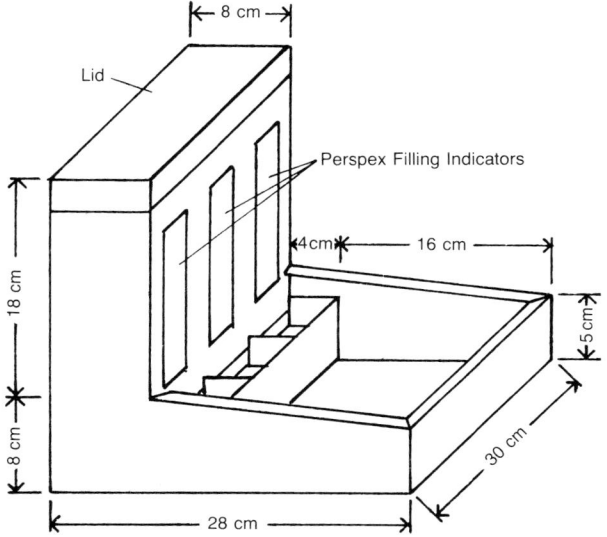

Figure 5. Combined seed hopper and feeding tray.

A supply of grit, cuttlebone and charcoal should be available at all times and should be supplied in a dish suspended above the floor, in a protected position.

Floor Surfaces

Dry, easy to clean floor surfaces are essential if a healthy environment is to be maintained in an aviary system.

Concrete floors sprinkled with sand provide a reasonably easy to clean surface, as the sand prevents droppings, etc., from adhering to the floor. Well drained concrete floors are easily cleaned by regular hosing.

Pebble or lucky stone floors which are provided with a suitable sub-surface drainage system are also easily cleaned by regular hosing with water.

Sand filled floors require a suitable sub-surface drainage system to keep them dry (see Figure 2) and may be cleaned by removing the soiled surface regularly. Total replacement of sand filled floors is usually necessary every year or two.

Earth floors require constant attention as they often become infested with vermin such as rats and mice as well as promoting worm infestation. Regular sweeping, raking and liming of the soil is advisable, particularly for the ground frequenting Neophemas.

Worming

All species of the Neophema group are vulnerable to worm infestation.

Neophemas housed in conventional aviaries, regardless of the type of floor surface, should be wormed at three monthly intervals.

Birds housed in suspended wire mesh aviaries should never need to be wormed providing the wire mesh floors are kept clean and free of excreta build ups.

Worming via the drinking water must be considered unreliable for Australian parrakeets as they are capable of not drinking for up to three days during mild weather conditions if they consider the water unpalatable. Even if they do drink it is impossible to monitor just how much of the worming agent is consumed.

The only sure and effective way to worm any psittacine is to catch up each bird and administer the worming agent direct to the crop, using a stainless steel crop needle attached to a syringe.

We normally use the worming agent "Panacur 2.5" which is a sheep drench marketed in Australia by Hoechst. The recommended dose rate for Neophemas is 0.3 ml direct to the crop. Panacur is not registered as a worming agent for birds, so the responsibility for its use must be accepted by the aviculturist concerned.

Recent research relating to worm infestations suggest it is advisable to change the type of worming agent periodically for continued optimum results. We advise "Nilverm Pig and Poultry Wormer" which is marketed by ICI in Australia, as an alternative. The recommended dose rate for Neophemas is 0.2 ml of a dilution of one part Nilverm to five parts water. Be warned, Nilverm is a toxic agent to birds — the five to one dilution and correct dose rate is essential. We have found this dose rate to be safe and effective in our hands but this drug does have a low safety margin. It is not registered for use in parrots and the aviculturist must accept responsibility for its use.

Occasionally death occurs after worming and is usually the result of an intestinal blockage caused by clusters of worms from a massive infestation. Infestation of this magnitude is rarely treated successfully. Panacur kills worms more slowly than Nilverm and it is the drug of choice in heavy infestations in an attempt to avoid impactions of dead worms.

The recent release (July 1992) of the product "Hapavet" marketed by Vetafarm and registered for use in parrots may provide a suitable worming medication administered via drinking water.

We make a practice of never worming birds on a hot day as losses have occurred in Australian Parrakeets from the stress of being wormed during temperatures in excess of 30°C.

Quarantine

All new birds brought into a collection must be quarantined for a minimum of 28 days in an area isolated from other birds. This area should be enclosed, well protected and draught free. The birds should be placed in easy to clean holding cages with false wire bottoms.

During the quarantine period the birds should be treated with chlor-tetracyclines or doxycyclines under veterinary supervision, wormed and checked thoroughly for any sign of disease.

After the quarantine period and when satisfied the new birds are healthy they may then be introduced into the collection.

Routine Daily Observation

The practice of observing all birds in your collection at least once and preferably twice a day is advisable. Any bird that appears inactive, listless or dull in the eye, should be caught up and examined for loss of weight, dirty vent, infections in the mouth or around the beak, sore eyes or injuries and the necessary measures taken. Prompt diagnosis and treatment is essential if a satisfactory recovery is to be achieved.

Any bird requiring treatment or observation should be placed in a cabinet, holding cage or small aviary, preferably situated in a bird room, garage or some other warm place, to simplify medication and further surveillance. Artificial heat should be applied when deemed necessary.

Sore Eyes

This group of small parrakeets are particularly subject to sore eyes caused by bacterial infections commonly referred to as conjunctivitis. In Neophemas these cases should be treated as Chlamydiosis (psittacosis) until proved otherwise. The symptoms start with a slight irritation in one eye which causes minor swelling of the eye lid and blinking. As the infection progresses the swelling increases, the feathers close to the eye fall out and the other eye becomes infected.

Unfortunately it is often not until the second eye is infected that the disease is actually detected, for during the period when only one eye is infected the bird will always look at you with its good eye, which means the infection usually has a good hold before treatment is commenced.

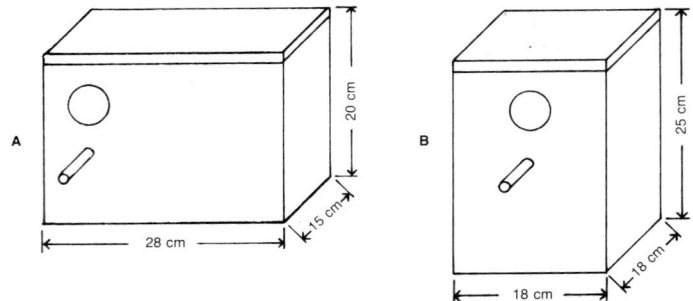

Horizontal and vertical commercially made nest boxes.

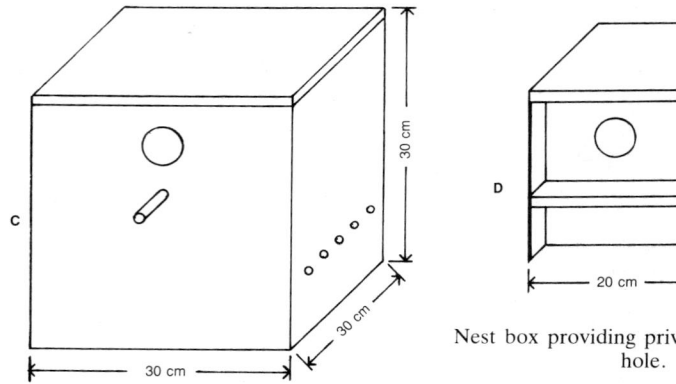

Large ventilated nest box suitable for Blue-wings.

Nest box providing privacy for entrance hole.

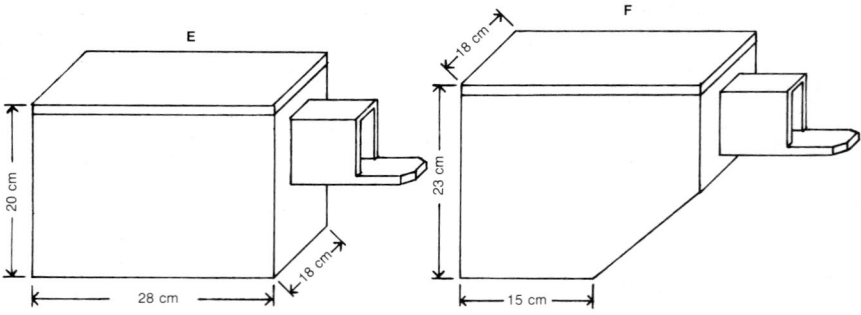

Nest box with entrance spout suitable for Rock Parrakeets.

Nest box favoured for all species by some breeders

Figure 6.

40

When a case of conjunctivitis is detected the bird concerned must be isolated, as it is an infectious disease, all other Neophema inmates of the aviary should be checked for signs of the infection, and all perches, ledges and other areas where birds may land, washed with distinfectant.

Conjunctivitis appears to be spread by infected birds wiping their sore eyes along perching areas.

In most cases the infection can be cured by treating the bird with chlortetracyclines or doxycycline in the drinking water for long time periods, as prescribed by your avian veterinarian. Severe infections may require a longer period of treatment and the sparing use of eye ointment.

The cause of conjunctivitis can usually be traced to contact with a bird that has been exposed to dirty or overcrowded conditions.

Our suggested quarantine procedure for new birds will help combat this disease.

Nest Facilities and their Cleaning

Suitable nest sites for Neophemas fall into two categories — hollow logs and nest boxes.

Hollow logs are the natural nesting site for all species except the Rock Parrakeet. The only advantages for hollow logs are durability and their acceptability by pairs refusing to breed in a nest box.

The disadvantages of hollow logs are their scarcity, restricted access, difficulty of inspection and the problems associated with thorough cleaning.

We have had this group accept a wide variety of hollow logs, both large and small, but the most convenient type and size is a log about 30 cm (12 in.) long with an internal diameter of 12 cm (5 in.) to 20 cm (8 in.), an entrance hole on the side near the top and a removable lid for access. Vertical or partially inclined logs are most suitable as the young are unable to reach the entrance hole until they are well developed.

Nest boxes have many advantages over logs when used for this domesticated group.

There is a large variety of nest boxes used for Neophemas in Australia (see Figure 6) with each design having some particular quality which endears it to individual breeders.

We find that on a whole this group is easily satisfied where nest sites are concerned so the features we look for are adequate size, good access and easy cleaning.

The commercially made nest boxes, which are inexpensive and adequate for most species, provide a viable method of disease control by discarding all used nest boxes at the end of each breeding season.

Nests should be filled to a depth of 2.5 cm (1 in.) to 5 cm (2 in.) with any of the following materials or combinations of them; rotting wood dirt from the inside of dead stumps or trees, damp fine wood shavings or peatmoss.

All nest sites should be cleaned and scrubbed with a disinfectant such as "Sani-Chick" prior to the breeding season and between each clutch when possible. Nests must also be sprayed with a suitable insecticide for lice and mite control such as "Duramitex", "Coopex" or a maldison based product such as "Malathon" which is a garden spray and an effective mite control agent when sprayed in a 50 part water to one part Malathon dilution. Malathon is produced by Chemspray Pty. Ltd. in Australia. These products are not registered for use with parrots and the aviculturist has to accept responsibility for their use. Bourke's Parrakeets appear to be especially susceptible to toxicity from Malathon if not diluted.

Extreme care is required when selecting a method of fixing a nest site to the aviary wall or ceiling. Never allow a wire end, nail, small screw, hook, double wire or any other conceivable object or circumstance which may be remotely possible to snag, hook or tangle a bird's leg band, leg or neck, for sooner or later the unlikely or even the impossible will happen.

The end result is usually a broken leg at the best, but more likely the loss of a leg or a slow agonising death.

Horizontal nest boxes should be hung at a slight angle to keep the eggs at the rear of the box.

Pairing

With the exception of the Rock Parrakeet, which we advise to have surgically sexed, all species in this group are visually sexed with ease.

Compatability of pairs in most instances is unimportant, although occasionally a pair is encountered that reject all attempts made to induce them to breed. Often such birds will breed when provided with different mates.

Preferably pairs should be placed together at least a month prior to the normal commencement of the breeding season.

42

Egg Breaking

Egg breaking is a vice seldom encountered in the Neophema group, but on those rare occasions is best addressed by providing a longer, partially inclined nest box or log. This slows the birds entry to the nest chamber, thus lessening the chance of egg damage as well as creating a more secure enviroment for the incubating hen.

Rearing chicks

We advise a daily inspection of the nest as soon as hatching occurs and wherever possible, twice daily. Some aviculturists do not agree with such interference but our experience indicates that the advantages gained by regular inspection far exceed the odd disadvantage that may arise. With daily inspections potential problems may be rectified before they become disasters.

For example, a chick that falls behind it's nest mates may be taken for fostering or hand-rearing while youngsters being poorly reared by their parents, or suffering from vitamin deficiencies, are detected before they reach the point of no return.

A suitable, varied and nutritious diet must be provided to pairs rearing young and care taken to ensure sufficient quantities are being fed by them to promote an adequate growth rate. If in doubt provide more variation to the diet until suitable rearing foods, which are more to the parents liking are found. See Diet Chapter.

Fostering

The use of foster parents within the Neophema group is widely practised for a variety of reasons, such as desertion of chicks; fostering small clutches into other nests to allow the pair to breed again; using reliable pairs to rear the young of unreliable pairs; or fostering the eggs or young of rare species or mutations to increase the production from these birds.

Overall, Neophemas are good foster parents and fostering can be practised with reasonable safety, from fresh eggs to hatchlings, to babies at any stage, providing the clutch into which they are being fostered is at a similar stage of development.

Cross fostering between species is also usually safe. We have cross fostered eggs and young successfully to fledging in various combinations and with multi-species clutches, using all species except the Orange-bellied Parrakeet. Neophemas have also reared members of Psephotus genus in our aviaries.

43

An interesting record of unlikely foster rearing is provided by Harvey Oliver of Wagga Wagga, New South Wales who had a pair of Cockatiels successfully rear a clutch of Bourke's Parrakeets to independence.

Regular inspection of nests containing fostered young is essential if early signs of favouritism in the feeding of individual babies or desertion by the surrogate parents are to be detected in time to relocate or hand rear the young.

Hand Rearing

There is probably less need to hand rear babies of the various species of this genus than in most genera of psittacines. Most Neophema species are totally domesticated and, on a whole, recognised as reliable parents, but the time always arises when hand rearing is the only option left.

If given a choice, the ideal time to take chicks for hand rearing is at the early pin feather stage. Chicks at this age will adapt quickly to their new environment and diet, require only three or four feeds a day and little artificial heat.

Avoid taking newly hatched chicks for hand rearing whenever possible, as chicks fed by their parents or foster parents for even just a few days have a much better start in life.

When freshly hatched chicks have to be hand reared we suggest the following procedures and diet which are the results of our recent research and testing of new techniques in "day one" diets.

We do suggest that unless good dexterity and eyesight are possessed and adequate time available, the hand rearing of newly hatched Neophema should not be attempted.

Our experience with this genus suggests that often the end result of hand reared chicks leaves much to be desired. This is especially the case if they are to be used as stock birds. There is certainly room for refining techniques and diets when it comes to Neophemas.

In accordance with our commitment to test and publish our findings on all new avicultural techniques, we suggest the following procedures and diets as suitable for all newly hatched Australian parrakeets, parrots and cockatoos, as well as most exotic psittacines.

Newly hatched chicks must be placed in a brooder which is capable of accurate temperature control and maintained at 97°F. If the brooder has humidity control about 30% humidity is adequate, otherwise a small, open container of water placed in the brooder will provide enough humidity to prevent dehydration of the chicks.

We will now describe the procedure of "crop washing" which is used to introduce the necessary bacteria into the hatchlings crops to ensure the satisfactory operation of the digestive system.

Select one of the parents, or any other healthy Neophema, and deposit 1.5 ml of warm, boiled water into the crop via a metal crop needle and syringe. Massage the crop gently to mix the contents with the water, allow to settle, then withdraw as much of the resulting fluid as possible with the crop needle and syringe.

When withdrawing the fluid from the crop, care must be taken to locate the end of the needle in such a position in the crop so as not to block the end of the needle with large pieces of half digested food or the wall of the crop.

This process has become known as "crop washing". Crop wash must be stored in an air-tight container under refrigeration and should not be kept for longer than three days. The procedure is not as difficult as it may seem — after a little practise most can become reasonably proficient.

Within two hours of hatching, dissolve a pinch of glucose powder into several drops of warm, boiled water and feed enough to the chick to slightly extend the crop, using a tiny feeding spoon (a small, cut down teaspoon or salt spoon with the sides folded up to form a shute), the open end of a syringe, eyedropper or any other preferred feeding device. Always scald feeding equipment before use. Repeat every two hours until the chick uses its bowel to pass the green faeces resulting from digesting the last of the yolk sac.

Once the green faeces have appeared the chick should then be fed the day one "baby diet" (as suggested in the Diet Chapter) diluted to a runny consistency with a few drops of "crop wash" and boiled water. Warm to body temperature by standing in a bowl of hot water. Feed only enough to slightly extend the crop. Extreme care must be taken not to overfeed as these tiny chicks are easily choked.

After two more similar feeds at two hourly intervals and providing good digestion continues, dilute two parts of "baby diet" with one part of "greenfood puree" (as described in Diet Chapter) plus a few drops of "crop wash". Feed this mixture at two hourly intervals from 6:00 a.m. to 10:00 p.m. with one nightly feed at 2:00 a.m. Continue to add "crop wash" to each feed until it runs out or is discarded after three days.

Providing all is still well, introduce a pinch of a dry mixture of equal parts "Farex" and ground "Egg and Biscuit" rearing food. If digestion continues for 24 hours replace the Farex and Egg and Biscuit mixture with a similar quantity of finely ground hand rearing food, as suggested in the Diet Chapter (Farex, Egg and Biscuit, Chicken starter crumbles and sunflower kernels).

At this stage the 2:00 a.m. feed can be discontinued and as the chick's crop capacity increases the feeds can be spaced to 2.5 hourly and then 3 hourly intervals.

Over the next few days gradually reduce the "day one baby food", retain the "greenfood puree" and increase the "hand rearing diet" accordingly. Then when the "day one diet" is totally replaced, and providing all still goes well, gradually discontinue the "greenfood puree" over a 24-hour period and continue with the "hand rearing diet" until the chick is weaned.

Weaning is a gradual process usually initiated by the chick itself. As it grows and the crop capacity increases, the number of feeds required each day decreases, eventually to only a morning and evening feed, when fully feathered and almost ready to fly.

At this stage the youngster reduces food intake dramatically so as to loose body weight, thus enabling an easier transition to flight. Now it should be placed in a small holding cage, which is situated indoors, with a variety of foods available such as dry seed, sprouted seeds, green peas, greenfood, milk arrowroot biscuit, plain cake, apple, etc. Hand feeding should now be reduced to once daily, in the evening to ensure the chick has food in its crop to sustain it during the night. When the chick is found to have about a small level teaspoon of food in its crop, hand feeding should cease and the chick's food intake monitored for a few days.

As soon as it is ascertained the youngster is feeding on a regular basis it should be removed to a holding cage in a birdroom or garage and then, if all is still well, after two weeks relocated to a small sheltered holding aviary.

Always remember when hand rearing that each progressive step must be taken slowly and be prepared to retrace a step if necessary. Monitoring of the chick's weight is a reliable guide to its progress.

Artificial Incubation

Artificial incubation of Neophema eggs should be resorted to only when no other options are available as the successful hatching rate of artificially incubated eggs is often less than desirable.

Fostering eggs under other members of this group or Budgerigars is a more viable alternative, even if only for the first two weeks of incubation as then hatchability will be improved considerably.

The following data and procedures are submitted as a general guide to artificial incubation and should be used in conjunction with the instructions supplied with the incubator being used.

Always thoroughly clean and disinfect the incubator before commencing operation, then monitor and make necessary adjustments during 24 hours of running prior to placing the eggs in the incubator.

The normal operating temperature for parrakeet eggs is 99°F although some aviculturists prefer temperatures 0.5°F higher or lower. A relative humidity of 55% is advisable although again some prefer a few percentage points higher or lower which are equivalent to wet bulb thermometer readings of 84°F to 88°F.

Ideally, before fresh eggs are placed in the incubator they should be lightly cleaned and their weight recorded so the weight loss incurred during incubation may be monitored. The generally accepted ideal weight loss of 16% during the whole incubation period should be aimed for.

This is calculated by dividing the initial weight of the egg by 6.25 to obtain the preferred 16% overall weight loss, which in turn should be divided by the number of days in the normal incubation period for the particular species, to ascertain the ideal daily weight loss. All this information should be recorded for each egg prior to it being placed in the incubator.

Eggs should be placed in the incubator racks, which may need to be adapted to accommodate small Neophema eggs, with the large blunt end uppermost and inclined at approximately 45° to the horizontal. Some automatic incubator racks hold and roll the eggs in a horizontal position, with good hatching results.

If the incubator has no automatic turning device, the eggs must be turned manually from a 45° inclination in one direction to 45° in the other, commencing 24 hours after the start of incubation. A minimum of three turns per day (morning, midday and evening) is essential and additional turning advantageous, but the eggs must always be turned an odd number of times each day so the eggs do not lay in the same position consecutively during the long night period.

Although automatic turners rotate the eggs regularly, some feel it is desirable to manually turn the inclined eggs from one direction to the other twice daily in addition to the automatic turning.

After three days the eggs should be weighed again and compared with the theoretical weight loss for a three-day period. If the eggs have lost too little weight, this suggests the humidity is too high and should be reduced. Conversely, if the eggs have lost too much weight, they are drying out too quickly, hence the humidity should be increased. This weight loss should be monitored closely and the weight checked every few days and then appropriate adjustments made to the humidity.

In practice we have found that eggs often hatch successfully with considerably more or less weight loss than the advised ideal of 16%.

Three days prior to the eggs hatching, the humidity should be increased to about 75% or 92°F on the wet bulb thermometer, and the eggs placed in a fine wire or plastic mesh basket which allows full air circulation around the eggs, yet protects the chicks when they hatch as well as stops the turning of the eggs. Some find they have increased hatching results when the eggs are placed in a separate incubator, often of the still air type, for the last three days of incubation.

We submit the preceding information as a "starting place" that, when supported with personal experience and a working knowledge of your own particular incubator, will lead to successful hatchings.

Currently we are using a "Rotarex" incubator which is made in New Zealand by Dominion Incubators Pty. Ltd. and distributed in New South Wales by Geoff Andrews Pty. Ltd. This incubator is a refined and more manageable version of the Marsh "RollX" and is providing us with our best hatching results to date.

Another viable use for artificial incubation is to provide a holding place for deserted eggs until fostering arrangements can be made. In other words an operational incubator provides those few days leeway so often needed.

48

Colony Breeding

Colony breeding is when a group of birds of the same species are housed in the same aviary for the purpose of reproduction.

There are numerous records of successful breeding colonies for most Neophema species, although some species are more suitable than others (see Species Chapters).

Generally speaking, we feel the disadvantages of the colony system, such as fighting and the stress created by dominant pairs within the colony, far exceed the one advantage of saving space.

The only fully successful colonies, where all pairs breed, that we are aware of, are housed in very large aviaries which allow each pair to establish a breeding territory. Even under these conditions the colonies are limited to five or six pair.

If a successful colony is ever established be careful not to upset the delicate balance; never add to or reduce the colony's numbers, remove the young as soon as they are independant and never relocate the group or change the facilities in the aviary.

Usually successful colonies end when the unique combination is changed in some minute or obscure way which tips the invisible balance.

Partially successful colonies are more common, where one or two dominant pairs breed and the remaining birds in the aviary just make up the number.

Flock Breeding

Our definition of flock breeding is where a group of dissimilar parrakeets, or birds generally, are housed together for breeding purposes.

Single pairs of Neophemas will breed readily when housed in flock situations with dissimilar birds such as finches, doves or smaller softbills.

There are some instances where pairs of more than one species of Neophemas housed with flocks of dissimilar birds breed well, although fighting between the pairs often eliminates breeding totally or with the exception of the dominant pair. The Neophemas will not disrupt the breeding of finches, doves or softbills they are housed with.

Likewise flocks of dissimilar Neophema species only, are rarely totally successful except when limited to two pairs and even then the results are dependant on the particular species housed together and the size of the aviary. The chances of success are increased if one of the species is the Bourke Parrakeet and the aviaries are large and heavily planted.

We have had reasonable results in breeding a pair of Neophemas housed with pairs of larger parrots such as Princess, Superbs, Regents, Kings, Crimson-wings and Kakarikis. Individual pairs of these parrots will tolerate one species of Neophema yet reject another. For example we have had pairs of Princess reject Turquoise Parrakeets and Elegant Parrakeets as co-tenants yet accept Scarlet-chested Parrakeets. Superbs have also shown a preference for particular species as co-tenants.

Transportation

As a general rule Neophemas are placid birds which show no aggression towards each other when subjected to the stress of close confinement during transportation. Yet even in this docile group there are exceptions to the rule.

Recently Stan had a bad experience when a friend brought an established mated pair of Turquoise Parrakeets to him in a carrying box of adequate size with a perforated solid front which provided a secure environment. When the pair was removed from the box after less than one hour, the male was found to have been scalped by the female — he subsequently died a week later despite all efforts to save him.

This female had never shown any signs of aggression towards her mate before, nor has she towards her new mate since, yet when placed under the stress of what appeared to be a perfect carrying box, she was transformed into a savage.

We have long advocated carrying or transportation boxes with separate compartments for each bird when conveying all the larger as well as the aggressive species of psittacines. Now, after the preceeding experience, we advise separate compartment boxes for all psittacines, regardless of the genera or the species, for sooner or later maiming will occur and when it does you can rest assured it will be the bird you could least afford to have killed or maimed.

Animal Welfare and Wildlife Authorities have still not provided adequate regulations or guidelines relating to the transportation of birds, particularly parrot species. Legislation in this area must only be a matter of time.

All species of the Neophema group should be transported in adequately ventilated, individually compartmented boxes, which should not be large enough to allow excessive fluttering that could lead to injury.

The following points should be considered when constructing transportation boxes for Neophemas:

1. Only lightweight construction is necessary for this group because of their limited chewing capabilities, and is also a consideration when the cost of the transport is related to weight.

   Three-ply sheeting is an ideal material. Boxes may be constructed with one or multiple compartments.

2. Smaller rather than larger compartments are advised to avoid injury by the thrashing about of frightened or stressed birds.

   Our formula for the minimum compartment size of a transport box is the overall length of the bird squared to calculate the base area with a height of the standing height of the bird plus 2.5 cm (1 in.) (see Figure 7).

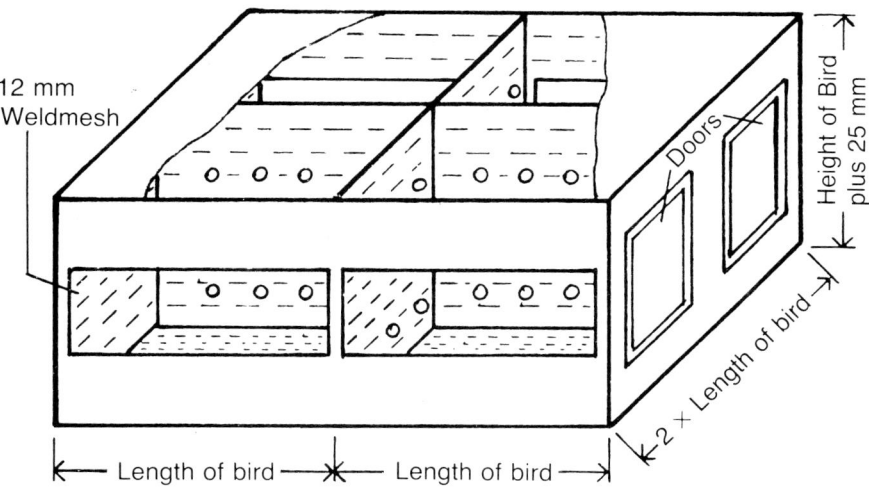

Figure 7. Multiple compartment transportation box.

3. The box should be totally enclosed with either a perforated front or a front consisting of one-third of a maximum 12 mm × 12 mm (½ in. × ½ in.) square wire mesh. The perforated or partial wire front provides the security required to alleviate fright and stress.

4. Ventilation holes should be provided in at least two sides of each compartment to allow free air circulation in the event of high temperatures being encountered.

5. When the duration of time to be spent in transit is under four or five hours it is not necessary to provide water for the trip, however, water containers with the top edge turned over to form a lip, thus preventing the water waving out, should be fixed inside the box for longer journeys.

6. Seed spread on the floor of the box is more satisfactory than placing it in a container where it may not be found or even ignored. A slice of apple provides a source of moisture when chewed by the birds.

7. Birds should never be shipped during heatwave conditions unless in air-conditioned vehicles.

8. When birds are to be transported unaccompanied by air, rail or road, the box should be labelled "LIVE BIRDS", correctly addressed and a contact telephone number stated, as well as provided with adequate food and water with 100% safety margin, for the duration of the trip. The recipient must be notified well in advance of the arrival time.

9. When a Wildlife Authority permit is required by law, it must be obtained prior to the transaction and a copy of the permit attached to the box.

DISEASES

James Gill, Bvsc, Mvm, Macvsc

Successful keeping and the subsequent reproduction of Neophemas is a function of the proper selection of good birds in the first place, followed by correct housing, good nutrition and management programmes including effective disease control.

Neophemas are the most domesticated genera of Australian parrots and therefore the possibilities of maximising production from them are greatly increased.

I have noticed for some time that with increasing levels of domestication there is a concurrent increase in reproductive performance. During the last fifteen years Bourke's Parrakeets have gone from being seasonal breeders to year round breeders. Many pairs of Scarlet-chested Parrakeets will now nest year round. Elegant Parrakeets would rarely double brood and they now regularly have three clutches.

This is partly a result of aviculturists (subconsciously in most cases) selecting for reproductive performance. Those birds that have adapted well to captivity breed well and hence provide the future breeding stock.

It is also a function with many desert species of birds that by simply providing the single most limiting factor to their survival and breeding — food — they reproduce readily.

Scientific evidence based on Mallard ducks has also demonstrated that domestication results in earlier nesting in the spring and a delay in the end of reproduction till later in the summer. There is also a decrease in the synchronisation between the male and female hormonal cycles. (Donham Biol. Reproduction, Vol. 21 Pages 1273–1285, 1979).

Neophemas as a group are subject to the same range of diseases as are other birds.

The wide range in monetary values of Neophemas often makes it difficult for owners and avian veterinarians to know to what extent to utilise veterinary services. It is always a good idea to have a clear understanding with your veterinarian what you hope to achieve with your birds.

53

Do you want to have the birds largely for beauty, and minimising mortality is more important than maximising reproduction and cost is of secondary importance?

Do you want to keep the birds for purely commercial reasons and the cost of veterinary services must be justified in terms of the overall profit of the collection?

Avian veterinary expertise has improved dramatically in the last fifteen years but unfortunately in a commercial operation the cost of saving a normal Bourke's Parrakeet may not be justified. However, if the Bourke's Parrakeet was a new Mutation it may well be worthwhile to get your avian veterinarians to use all methods at their disposal to save the bird. Likewise if the normal Bourke's Parrakeet was part of a large collection it may be worthwhile to have a full autopsy done to safeguard the rest of the collection.

These are considerations that all aviculturists must contemplate when trying to make decisions on how best to utilise their avian veterinarian. A good understanding and line of communication with your veterinarian is essential for this to work well.

Any bird can get one of two forms of disease — infectious or non-infectious disease. For birds to become ill with an infectious disease they must be exposed to a sufficient number of the infective agent (for example, virus particles or worm eggs) to overcome the bird's natural defence mechanisms, that is its immune system. Non-infectious diseases include such things as fractures of the legs and wings, egg binding, starvation, vitamin and mineral deficiencies. Some of the non-infectious diseases can predispose the birds to infectious diseases. The strength of the bird's immune system can be lowered by such things as cold damp conditions, deficiencies in vitamins and minerals, lack of sufficient quantities of food to eat and constant stress from other birds, dogs, cats, mice, rats or people, including their owners.

Therefore to minimise the chances of birds getting sick we must:

1. Reduce the bird's access to bacteria, viruses, parasites, etc. This can best be done by a month's quarantine on all birds coming into the collection and by maintaining good hygiene for all birds in the collection.

2. Make sure the birds' immune systems are working well. This is best done by ensuring that they are housed and fed well.

For those with a mathematical mind the chance of a bird getting an infection can be expressed as a formula:

$$\text{Infection} = \frac{\text{Numbers of pathogen} \times \text{type of pathogen}}{\text{bird's immune system}}$$

We should all strive to make preventative medicine the more important aim in our hobby and curative medicine a secondary role.

A large percentage of disease problems in any livestock system can be eliminated by good management. The system of housing, feeding and management as outlined by the authors in this book will go a long way towards eliminating disease problems.

Proper health management is at least partly a function of the proper planning, design, construction and maintenance of the aviaries themselves. Several important features should be highlighted here as they relate to disease control.

It is essential that fresh water be provided daily. For the aviculturist who has more than a few aviaries, to do this consistently an automatic watering system is an advantage (see Housing Chapter).

Concrete floors are considered by me to be essential. They enable aviaries to be cleaned thoroughly. To assist in this routine cleaning and drying, they should slope gently towards a drainage system. Walls should be of single thickness construction so as not to provide hiding places for mice or other vermin.

There should always be two doors between the birds and the outside of the aviary so as to prevent unnecessary losses due to escapes.

The divisions between aviaries should be of solid material or double wire to prevent traumatic injuries from neighbouring birds.

Perches should not be placed over food or water containers.

Food and water containers should be easily cleanable.

All new wire should be weathered or treated with vinegar and a wire brush to remove excess galvanising and oxide. Heavy metal poisoning from new wire is a common yet easily avoidable disease.

Remember all that any aviary has to do is provide a secure hygienic environment that gives the parrakeet the ability to exercise. Many other adornments are usually only for human aesthetic needs.

Quarantine

The isolation or quarantine of all incoming Neophemas for a period of at least 30 days is the first step in a successful disease control programme. This period enables the proper examination of the birds and observation of their behavioural patterns particularly their eating habits.

During this period the birds should be examined for any deformities, if they were not examined prior to purchase. They should be treated for any external and internal parasites (see Management Chapter).

The most important feature of quarantine is to prevent the introduction of new infectious agents into the collection. No bird should be removed from quarantine area to the general collection unless the aviculturist is completely happy with its health.

With valuable birds or valuable collections it is prudent to weigh the birds regularly during quarantine and to do direct faecal smears and faecal flotations at least twice looking for internal parasites.

With the legal importation of avian genetic material into Australia it would also be wise for aviculturists to quarantine birds once they leave the government quarantine facility. These birds may be introducing new infectious diseases that are not tested for by the government authorities. I would recommend the use of sentinel birds in your private quarantine areas as a simple effective way of checking for any potential problems. The cockatiel would be ideal for this purpose alongside any parrot. Any bird be it a new arrival or sentinel bird that dies in your quarantine should have a thorough post mortem performed on it.

Another major advantage of quarantine, which is often ignored, is that it provides an observation period to note the bird's behavioural patterns particularly their eating habits. Many newly acquired birds, especially freshly trapped wild birds, die of starvation, because they are not familar with the food offered and refuse to eat it. By keeping newly acquired birds by themselves or in pairs you can observe if they are eating the diet provided and adjust the diet accordingly. It can be lifesaving to sprinkle some seed on the floor of the cage or aviary until the birds find the seed dish.

The quarantine area should be isolated from the other aviaries and the birds should be fed and watered after the other birds in the collection have been cared for.

Diseases of the Skin and Feathers.

Traumatic injuries to the skin, beak and toes are quite common but should be largely prevented by good management. Most wounds resolve without complications but particularly toe injuries require vigorous antibiotic therapy to prevent the spread of infection. Beak injuries can be quite complicated and have a devastating effect on the bird especially if they loose the top beak. Some of these injuries require surgery which in my hands at least has a guarded prognosis. It is surprising how many parrots learn to eat without a top beak and providing they can survive the first few days do very well. Bleeding can be a problem with some of these injuries and is best controlled by restraining the bird and applying direct physical pressure to the area for two or three minutes. If this is not successful Condys Crystals (Potassium Permanganate) can be applied to small wounds to achieve haemostasis. Severe wounds will require veterinary attention.

Bumblefoot is an infection of the sole of the foot of parrots. It is particularly common in pet birds housed in unhygienic cages with ill sized perches and fed faulty diets. Many of the milder cases respond well to antibiotic and injectable Vitamin A therapy. The Vitamin A appears to improve the health of the tissues in the foot. The severe cases have a poor prognosis for complete recovery.

Mutilation of feathers can occur in a number of ways. The most common form is self mutilation, where birds, through boredom, sexual frustration or an irritation of the skin or deeper parts of body, start chewing their own feathers. You should, with the help of your avian veterinarian, try and find the cause and remove it. With breeding birds we routinely pluck the damaged feathers and adjust the environment to try and overcome the boredom. In pet birds if modification of the environment is unsuccessful then the use of injections of Medroxyprogesterone Acetate by experienced avian veterinarians has proved to be successful in a large percentage of cases. Some adult pairs will also pluck one another. Thankfully Neophemas as a group do not have the degree of trouble that is seen in some other groups.

Parents will also pluck their chicks in the nest and in severe cases will damage wings and toes in the process. The simplest solution to this problem is to remove the young for hand rearing.

Psittacine Beak and Feather Disease is one of the most common and frustrating diseases facing keepers of parrots. In our practice in Sydney we see large numbers of Sulphur Crested Cockatoos affected with this disease in there first year of life. Most of these birds have been illegally taken from the wild as nestlings. Smaller numbers of the other species including Neophemas are also seen with the disease including aviary bred birds. Drs Pass and Wylie, working at Murdoch University in Western Australia, have succeeded in transmitting the disease by inoculation of susceptible young birds with the virus particles. Continued research is being carried out both in Australia and overseas. Hopefully testing procedures for detecting carrier birds will be readily available in the near future.

Internationally there appears to be differences in the pathology and aetiology of the disease. It may well be that different aetiologies produce the same clinical picture. In time this confusion will no doubt be corrected as our understanding of the disease process in avian skin and feathers improves.

The features I look for are pinching or narrowing of the feather shaft as it leaves the feather follicle. This may be accompanied by haemorrhage in the feather shaft. These feathers are therefore brittle and fall out easily and the bird can end up completely naked. The birds are often very dirty in appearance. Skin biopsies may be necessary to confirm the diagnosis in mild cases. Deformities of the beak (beak rot) are a common feature of the disease in Sulphur Crested Cockatoos and Gang Gangs but not in Neophemas. In Neophemas one of the prominent features is a change from green to yellow or from blue to white feather colour.

It is an infectious disease and affected birds should be removed from a breeding collection.

Diseases Associated with Reproduction

Egg binding is a common problem with a complex aetiology (cause). Calcium deficient diets, obesity, chilling, lack of exercise and infections of the oviduct can all lead to egg binding. I find the majority of cases respond well to an increase in environmental temperature and administration orally of calcium (Calcium Sandoz Syrup R 0.2 ml every two hours for Neophemas until the egg is passed). In very depressed hens full supportive therapy is instituted, including intravenous fluid therapy. In birds that fail to respond to this regime and who are deteriorating, surgical therapy is considered. Radiographs may be needed to confirm the diagnosis. We are seeing more cases in our practice as it becomes more common for Neophemas to breed during the winter months. Cold weather can predispose to egg binding.

Fortunately infertility is not a major problem with any of the Neophemas. Most problems associated with failure to breed can be traced to management problems related to housing, nutrition and incompatiblity of the breeding pairs. Full clinical workups, including semen evaluation and endoscopic examination of the gonads in the breeding season, are essential in evaluating non-breeding birds. The use of mammalian hormones in birds in a blind attempt to stimulate breeding is of doubtful value and may be dangerous in the long term. Research has shown that even though one particular hormone (PMSG) resulted in stimulation of ovaries in some species of birds, it did not produce normal eggs.

I personally consider nutrition to be the single most important limiting factor in the reproductive performance of parrots in captivity. In recent years tremendous advances have been made in our understanding of housing, management and disease control in captive birds but unfortunately our knowledge of the dietary requirements of the many species of parrots is lagging behind.

Generalised Infectious Diseases

A wide range of bacteria have been reported to be pathogenic in Neophemas. Most of these are best controlled by good hygiene and quarantine. When a bird becomes infected it generally involves more than one body system at a time. If the illness is detected early, with improving treatment regimes including injectable antibiotic and fluid therapy many birds can be saved. With valuable individual birds or when outbreaks occur in a collection then bacterial culture and sensitivity testing is essential to maximise the chances of successful treatment.

It should be remembered that some of these bacterial diseases, most noteably Salmonella and Chlamydia, are potentially dangerous to the human owners of the birds.

One of the human tests (Clearview) for chlamydia has been adapted for use in birds by Australian veterinarians and is giving good results in confirming this disease in the live bird.

Chlamydiosis is a major cause of mortality in Neophemas particularly during the autumn when large numbers of birds pass through Bird Dealers shops. In fairness to the dealers they are at the mercy of what people sell to them. All newly acquired Neophemas should go on to a preventative treatment of chlortetracyclines or doxycyclines whilst in quarantine.

All bacterial diseases require experienced veterinary treatment to maximise the chances of a successful outcome.

Nephritis or infections of the kidney are a common primary disease especially those kept in unhygienic conditions. Bacteria from dirty perches invade abrasions or small cuts in the bird's feet and can be picked up by the blood supply and deposited in the kidneys. The birds become ill very quickly and are often vomiting and have profuse urine output. Depending on the degree of damage done to the kidneys they respond well to antibiotic and fluid therapy. I routinely recommend to owners of pet birds that they have two sets of perches and rotate them week about and clean the set not in use during the week and leave them out in the sun to dry.

Fungal Diseases

Aspergillosis is generally a chronic disease in parrots involving the respiratory tract, especially the lungs and air sacs. It is not a common problem in southern Australia. Candidia can become a problem of the upper digestive tract of nestlings. We routinely use Mycostatin in hand reared parrots that have any slowing of food passage time or in babies that start vomiting. The fungal elements can be seen on stained samples of the crop contents.

Parasitic Diseases

Ascarids or roundworms are a major problem in all parrot species. They can be controlled by the use of concrete or wire floors and by worming all birds as they come into quarantine. In Australia Panacur 2.5 (Fenbendazole 25 g/l) is the most commonly used worming drug for Neophemas. It is not registered for use in birds and the user must assume responsibility for its use. The following dose has been used with safety, 0.3 ml/Neophema. It should not be used when birds are moulting as it has been incriminated in causing feather abnormalities.

Capillaria have not been a major problem in Neophemas in our practice. However, they are more difficult to erradicate. I use Panacur 2.5 at 0.2 ml/100 gm bodyweight daily for three days.

Scaly Face Mite (Cnemidocoptes mites) are a major problem in Scarlet-chested Parrakeets. As well as the classical form around the face and feet, this particular species also gets a generalised form involving all of the body and the primary wing and tail feathers. The use of Ivermectin at 200 ug/kg has been found to be effective. Repeated dosing at three week intervals for three treatments is often necessary. Get your avian veterinarian to prepare the diluted mixture of the drug so as to ensure the correct dose rate is given.

Surgical Sexing

Surgical sexing is essential to get pairs of Rock Parrakeets. It is relatively safe, accurate and a fast way of determining the sex of these birds.

TURQUOISE PARRAKEET

Plate 1. Turquoise Parrakeets (pair) male on right.

TURQUOISE PARRAKEET

(Neophema pulchella)

Derivation:
Neophema:
   Neo — from *Neos,*
      Greek for new
   Phema — from *pheme,*
      Greek for voice
pulchella — from *pulchellus,*
   Latin for pretty

23° 27'

Description:

   Length 20 cm average
   Weight 40 g average
   Appearance as in Plate No. 1

Classification

The Turquoise Parrakeet, *Neophema pulchella,* is one of the two "type" members of the *Neophema* genus which are the smallest of this group of small ground frequenting parrakeets. Both have bright green body colouration, full blue faces and wing bands and pronounced sexual dimorphism.

These features make the typical Neophemas quite distinct from the Neonanodes (blue-browed) group of this genus.

Gregory Mathews cited a slight difference in the beak formation of the Neonanodes group in comparison to these species. He also separated the Victorian race as *N. dombraini* which was said to have more red on the wing of the male and less blue. The subspecies is not currently recognised.

Earliest Report

The Turquoise Parrakeet was first described by Dr. G. Shaw in *The Naturalist's Miscellany* Volume 3, Plate 96, 1792 as *Psittacus pulchellus.*

In 1891 Salvadori listed this species as *Neophema pulchella* in the *Catalogue of Birds in the British Museum* Volume 20, page 569.

The first published illustration appears to be that by Nodder, also in Shaw's *The Naturalist's Miscellany* Volume 3, Plate 96, 1792.

Range, Habitat and Field Notes

The Turquoise Parrakeet is a non-migratory species subject to local and seasonal movements which are dictated to a certain extent, by rainfall.

Historically, this once common species enjoyed a more extensive range which extended from the Dandenong Ranges near Melbourne in Victoria, through the eastern half of New South Wales to beyond the Tropic of Capricorn in Queensland where Gould's collector, John Gilbert recorded them in 1845.

By the turn of the century this parrakeet had suffered a massive decline in numbers throughout its range and was considered a rare species. The decline continued to such an extent that Gregory Mathews in his *The Birds of Australia* Volume 6, page 461, 1916 considered this bird to be extinct in Victoria and New South Wales.

From 1920 onwards recorded sightings of small numbers of Turquoise Parrakeets started to trickle through the ornithological literature and by 1950 regular sightings were recorded throughout its former range in New South Wales — even on the outskirts of Sydney.

The Turquoise Parrakeet had done the impossible — it had come back from the dead, and today has re-established itself in good numbers throughout much of its former range.

Currently this species' range extends from the Bendigo district of Victoria, eastward to the Mallacoota region, then north through the eastern half of New South Wales as far west as the Griffith, Broken Range and Marra Creek districts and then through south-east Queensland as far west as the vicinity of St. George and Chinchilla and north as far as the Maryborough region.

This parrakeet is found in open forest; along the grassed edges of eucalypt woodland; in forest clearings; up and around timbered slopes, ridges, gullies and watercourses which are adjacent to grasslands and farm-lands.

Turquoise Parrakeets always feed on or close to the ground.

The diet consists of the seeds of a wide range of introduced and native grasses, herbaceous plants and small shrubs, as well as vegetable matter and small berries. Amongst the wide variety of seeds, buds and berries they have been recorded eating are chickweed, paspalum, barley grass, wild mustard, stinging nettle, saffron thistle, grevillea alpina, bearded heath and wild millet.

This species is never found far from a permanent water source and always approach the water cautiously, observing the drinking place from nearby trees, before flying down to drink. On occasions they have been observed drinking before first light of a morning.

Sightings of Turquoise Parrakeets are usually of pairs or small groups feeding inconspicuously on the ground. When approached they are surprisingly unafraid and will try to walk away while continuing to feed. If pressed further they will fly low for a few metres then land and resume their feeding but if startled they will fly to the nearest trees, often uttering their soft, two syllable, whistling alarm notes. After satisfying themselves that the threat has passed they will float down to feed again.

Stan's first sighting of this species was of a lone female in mountain bushland at Hilltop (near Mittagong) on the Southern Highlands of New South Wales during the spring of 1962. He had located a Kookaburra's nest in a hollow in the trunk of a eucalypt about 7 m (23 ft) above the ground which stood on the edge of a small clearing. The parents were feeding their young on incredibly large earth worms which were as thick as a fore-finger — they would fly to the entrance of the nest and feed the worm to the eager chick. The Kookaburras were collecting the worms some distance away for each parent would take 15 to 20 minutes to return with a worm.

Settled back comfortably against a tree, Stan was waiting for the next Kookaburra to return when he noticed a small parrot fly into the clearing and cling to the fire ravaged trunk of an old dead eucalypt. Closer observation revealed a hen Turquoise Parrakeet clinging to the vertical burnt trunk nibbling the charcoal. She remained there for almost half an hour before flying back to the bush. Obviously this hen was rearing young and consumed the charcoal to feed to its chicks.

Another sighting with a humourous twist occurred a year or two later. Stan had some New Zealand visitors who were keen to see Australian parrots in the field. After a couple of successful birdwatching excursions into the mountain country west and south-west of Sydney, Stan decided to look for Turquoise Parrakeets for his visitors, where he had often seen them, along the Putty Road which runs from Windsor to Singleton, north-west of Sydney.

Having left Sydney early on the Saturday morning, Stan and Jill, their daughters Ann and Stephanie (nine and seven years old at the time) and their New Zealand friends were about half way along the Putty Road when a small flock of Turquoisines were seen to leave the roadside and fly to some small eucalypts one hundred metres away. Stan stopped the car and all vacated and stood still. Soon the Turquoisines floated down to feed again and provided good viewing for all.

By this time Ann and Stephanie, like little girls and not so little girls everywhere, needed to "spend a penny" and wandered down a dry creek bed so their modesty could be satisfied. Soon there were screams from the creek bed and the girls appeared at full gallop, trying to hitch up their knickers and slacks as they ran — with all thoughts of modesty forgotten in their terror. Stephanie was several paces ahead of Ann who was gasping out enquiries of what they were running from — she had no idea. When they reached the safety of their mother, Stephanie blurted out her story. Apparently she had just settled herself down in a secluded spot and gazed casually at the creek bank around her, when she looked straight into the eyes of a snake nestled into the bank only two metres away. Of course she took off and Ann, seeing her terror, followed in quick pursuit, without any idea of what she was running from. The girl's toilet stops were far less frequent for the remainder of that trip.

The flight of this parrakeet is fast, floating and irregular when on long flights whereas shorter distances are covered slowly in a floating, fluttering manner. When coming into land the tail is fanned to act as a brake.

Breeding in the Wild

Breeding usually occurs from August to December, being earlier at the northern limits of the range and later in the south. Autumn nestings have been recorded and probably relates to unseasonal rains. This species is probably double brooded in good seasons.

Nest sites are hollows in stumps, small trees, fallen dead timber, fence posts and dead trees, often close to the ground and seldom more than 2 m (6 ft 6 in.) high although nests have been recorded up to 9 m (30 ft) high.

Three to six white, rounded, oval eggs are laid on a base of rotten wood dirt and incubated by the female for approximately 18 days. Some incubating females have been observed carrying green leaves into the nest, by tucking them into the rump feathers in a manner similar to Peach-faced Lovebirds. The male feeds the incubating female either at the nest site or in a nearby tree during the morning and again in the late afternoon.

At first the young are brooded and fed only by the female, later as the chicks grow both parents share the duty. The young fledge after approximately 30 days in the nest.

After the young leave the nest the family parties move to feeding grounds where sometimes small flocks are formed.

Juveniles moult into adult plumage at about four months old.

Aviculture

Turquoise Parrakeets were bred as early as the 1860's in Europe and England. W. J. Greene in his *Parrots in Captivity* Volume 1, page 78, 1884 wrote "As far back as 1861 the Turquoise was bred in Germany and in Belgium; and in London Zoological Gardens some of these birds have been bred almost every season for many years back".

David Seth-Smith in his *Parrakeets* 1903, page 225, stated "The Turquoisine is at the present time extremely rare in this country," (England) then goes on to say "it has bred in captivity in this country and on the Continent on numerous occasions. In the London Zoological Gardens alone numbers were bred between the years 1860 and 1883."

Neunziz stated this species was first bred in Germany at Antwerp Zoo in 1861.

There are numerous references in literature to this species in captivity in Australia from the 1880's onward yet we can find no mention of successful breedings. Such a prolific parrakeet must have bred during this period. By 1915 Turquoisines were thought to be extinct in the wild as well as in captivity.

Alan Lendon in his *Australian Parrots in Captivity* 1951, page 74, states "A few of this species reached the hands of experienced Sydney aviculturists in the early 1920's, and since that date the species has steadily increased its numbers in captivity in most parts of Australia."

All early breedings of this species in Australia passed unrecorded until G. A. Heuman of Sydney, New South Wales, wrote the article "The Turquoise Parrot" in the *Emu*, Volume 27, page 68, 1927 in which he states; "As an aviary bird the Turquoisine is undoubtedly one of the most desirable. With other smaller birds it lives in peace, and even with other Parrakeets it is never aggressive and breeds quite readily in their company."

"I have caged Turquoise Parrots for many years now and seldom missed a year in their breeding. My aviaries are very large 'flights', but a friend who has a pair of my young birds bred them in a packing-case. They use a small box with a round opening and lay generally four eggs, all four being hatched as a rule. I have never been able to get more than one brood during the season, but my friend, Mr. Harvey Jnr. of Adelaide, to whom I sent a pair, obtained two broods of two pairs each, making four pairs for the season."

This first recorded breeding of the Turquoise Parrakeet in Australia by G. A. Heuman of Sydney, New South Wales, had to occur prior to 1925.

Another interesting snippet of information gleaned from this article was that G. A. Heuman had recently paid £30.00 ($60.00) for a trapped cock Turquoise Parrakeet. This must have been an incredible amount of money to pay for a bird in the 1920's.

The first recorded breeding in South Australia was by Mr. S. Harvey of Adelaide, South Australia in 1929. He was awarded the medal of the Avicultural Society of South Australia for this achievement.

Today the Turquoise Parrakeet is totally domesticated with many strains being aviary bred for scores of generations. The appearance of mutations in this species, particularly the beautiful "dilute yellow", increased its popularity as an aviary bird even further. They are so numerous in aviculture now that they may be purchased inexpensively — a desirable situation which relieves the pressure of illegal trapping on wild populations.

Stan first bred this species in 1960 from a pair housed in an aviary 3.6 m (12 ft) long, 90 cm (3 ft) wide and 2 m (6 ft 6 in.) high, the back half was a fully enclosed shelter and the front half was open flight. The pair nested in a small hollow log and fledged three young.

In 1967 Stan had a 12-month-old pair, which were descendants of his original stock, fledge eight young from their first nest, and five years on it was still the only nest of young that pair had produced. The female never did lay another egg. This is the largest number of parent-reared young, produced in the same nest without fostering that we have recorded for this species.

During the 1970's Stan became involved in the redevelopment of the Yellow Turquoisine. This mutation, which is recessive, had done reasonably well during the early years of its establishment but then reached a stage where they became poor breeders. A number of factors seemed to be involved, including a general weakness of the mutation, poor fertility and failure to rear their young.

Stan had the opportunity to purchase six of the few remaining Yellow Turquoisines in New South Wales — five old, well tried hens and one young cock. Convinced that the cure for this mutation's problems was the introduction of new blood (i.e., add to the gene pool), six fine specimens of aviary bred, totally normal (green) Turquoisines were selected to be used for outcrossing. Particular care was taken not to select "Red-Fronted" stock or birds related to this strain because of the constant inbreeding practised to retain and increase the red front on these colour variants.

The first season only three of the yellow birds bred, resulting in ten young splits being reared, two of which died before maturity.

For the second season three pair of the young splits were mated together and the remaining two mated to yellow birds. The result was four fine young yellow birds, 12 splits and 15 possible splits all of which were given to friends who eventually produced yellow birds from some of the pair combinations. Two of the original hens died during this season.

Due to the increased strength of the outcrossed stock used in the third season, 12 yellows, over 40 splits and numerous possible splits were produced.

From then on it was easy going, the outcrossed birds were prolific and in excess of 30 yellows and numerous splits were bred during the fourth season and similar results were achieved for the several years following.

Yellow to Yellow matings were now possible and the resulting young were superb. Whatever the breeding project, be it for type, colour mutation, size, etc., never neglect to introduce new blood.

The housing used throughout this project was a block of 15 aviaries, each measuring 2.15 m (7 ft) long, 90 cm (3 ft) wide and 2 m (6ft 6 in.) high, which were fully roofed, had solid partitions, concrete floors and were open only at the front. All the aviaries faced north and opened into a 1.2 m (4 ft) wide safety corridor.

Turquoise Parrakeets have been bred successfully in every conceivable type of housing, including conventional aviaries in a variety of shapes and sizes, large planted aviaries, suspended wire aviaries and small breeding cabinets.

We have found them to be inoffensive when housed with most other species, including finches and small softbills. Some pairs are intolerant of their own kind as well as other Neophema species.

In our own aviaries we have successfully flock bred Turquoisines with finches, softbills, pigeons and doves as well as inoffensive Australian Parrots such as Kings, Princess, Superbs, Regents, Cockatiels and Bourkes.

The optimum breeding results, in the long term, are obtained from pairs housed individually yet there has been some outstanding successes resulting from colonies housed in large aviaries.

The most successful colony we are aware of is that of Colin Cleak, a very experienced aviculturist from Shepparton, Victoria.

Six Yellow cock Turquoisines are housed with up to eight or nine Yellow or split hens in an aviary 11 m (36 ft) long, 6.5 m (21 ft) wide and 3.5 m (11 ft 6 in.) high, which is heavily planted with Melaleuca shrubs. There is a 3 m (10 ft) wide, fully enclosed shelter across the southern end which has fibreglass sheeting protecting half of the front and the western side is totally walled to provide full wind protection.

Approximately eleven nesting logs are hung both in the shelter and the open flight but the logs in the flight area are positioned so as to be protected from the direct sun by the shrubs. Some pairs change their log after each brood.

Most pairs are at least double brooded and the odd females are often serviced by one of the males and then rear their young unaided. Colin has flown between 40 and 50 young per season from this colony for the past six years.

In Australia, Turquoisines have not been renowned for their successes in breeding cabinets, yet in the late 1960's Stan's friend, Harvey Oliver, then of Sydney, now of Wagga, New South Wales, had remarkable success with a pair housed in a cabinet.

This cabinet was 1.8 m (6 ft) long, 50 cm (1 ft 8 in.) high and 45 cm (1 ft 6 in.) deep, its only protection from the weather was a metal roof fitted onto the cabinet itself which was located within an open service corridor.

A budgerigar nest box placed on the floor of the cabinet was accepted by the pair. The pair was housed in this cabinet for three years and during this period they nested three times each season and fledged a minimum of three and maximum of six young but usually four or five per nest.

Sexing

The sexing of adult birds is obvious with the male having a brighter blue face and red wing patches which are absent on the female except on very rare occassions when an odd red feather or a rusty mark may be visible.

Juveniles are similar to adult females with most young males being identifiable by at least a trace of the red wing patch when they fledge although on odd occasions this patch may not be visible until three or four months old.

Display

The male approaches the female stretched to his full height with the shoulders of the wing slightly spread, whilst uttering an almost inaudible whistling twitter.

Nests

Turquoise Parrakeets are not difficult to please with nest sites. A vast array of nest boxes are used for this species and providing they afford easy access for inspection and cleaning, they can all be deemed suitable.

We normally use the inexpensive, mass produced, horizontal or vertical "African Lovebird" type boxes which are readily available in pet shops (see Fig. 6A and 6B). Entrance spouts are not necessary for this species.

Hollow logs approximately 30 cm (1 ft) long and 13 cm (5 in.) in diameter with a lid on the upper end for easy access, the entrance hole near the top and hung either vertically or partially inclined provide ideal nest sites, particularly for that occasional pair that rejects nest boxes.

Any of the recommended nest fillings are suitable for this species.

Nesting and Hatching

This species will often have an extended breeding season when suitable weather conditions prevail. We have recorded eggs being laid as early as August 12th and as late as March 10th, although the vast majority of eggs are laid during September, October and November.

Just prior to nesting the female frequents the nest site with the male in attendance. Three to five rounded oval, white eggs (usually four but occasionally six to eight) are laid at two and sometimes three day intervals. Incubation is carried out by the female and usually commences with the laying of the second or third egg.

We have recorded incubation periods of 19, 20 and 21 days with the majority of eggs hatching in 20 days.

The chicks hatch with a silvery white down and at seven days old the eyes start to open, tail and flight pin feather development is prominent, as is

71

A Newly hatched nestlings.

B Ten-day-old nestlings.

C Twenty-day-old nestlings.

D Twenty-seven-day-old nestlings just prior to fledging.

Plate 2. Turquoise Parrakeet — nestlings.

72

A Dilute Yellow pair.

B Recessive Cinnamon male.

D Recessive Cinnamon (Plum-eyed).

C New Red-eyed Recessive Cinnamon male.

E Fallow. Note slight dilution and red eyes.

Plate 3. Turquoise Parrakeet — mutations.

F Red-fronted pair.

the grey secondary body down. At 14 days old all pin feathering is well advanced and at 21 days the chicks are almost fully feathered. We have recorded varying fledging periods of 24 to 34 days.

Fledglings are similar to adult females only with yellow beaks and with most young males showing traces of the red wing patch. At ten weeks old the beak colour has changed to brown. Adult plumage of the female and sub-adult plumage of the male is attained from four to six months old, while the brighter adult plumage of the male moults in during its second year.

Sexual maturity is usually reached the first breeding season after hatching, generally about 12 months old. Stan had a five months old hen produce fertile eggs when mated to a 12-month-old cock. Two chicks were fledged from the clutch.

Mutations

Australian Primary Mutations

DILUTE YELLOW See Plate No. 3A

This beautiful mutation was first bred by Bob Bush of Oatley, New South Wales, in 1961, from a pair of normals which had been bred by Syd Cook of Gymea, New South Wales. The young pair produced two yellow hens and a normal coloured cock in their first nest. The two yellow hens were acquired by the late Sir Edward Hallstrom of Sydney, New South Wales.

Stan saw these hens in Sir Edward's aviaries not long afterwards and was amazed to find both hens paired to Scarlet-chested males and each pair housed in "Macaw" aviaries about 12 m (40 ft) long, 3 m (10 ft) wide and 3 m (10 ft) high. Both hens died before they bred.

The pair which produced the yellows were sold to Frank Parmenter of Beverly Hills, New South Wales, and from this pair Frank was able to establish this recessive mutation.

Some difficulties were encountered during the early years of development, as the mutants appeared to be weak but when new blood lines were introduced the mutation strengthened and soon became well established.

This mutation is now well established throughout the world from birds taken illegally from Australia during the 1960's and 1970's.

Red-bellied or Red-fronted Yellow Turquoisines have been developed by line breeding with normal Red-bellied Turquoisines (See Plate No. 4E).

A Single-dark Factor and Normal males.

B Grey-green (double-dark factor) male.

C Normal and single-dark factor Dilute Yellow.

D Double-dark factor Dilute Yellow male.

E Red-fronted Dilute Yellow. The red on the female seldom extends beyond the belly.

Plate 4. Turquoise Parrakeet — mutations.

75

BLUE

The Blue mutation of this species first appeared in Sydney, New South Wales, in about 1985 and was said to have been produced from a Yellow hen mated to a split to yellow, normal coloured cock. Jim observed this mutation and assessed it as a true "Blue", lacking all yellow colour pigment. Good numbers were being built up but a set back in 1990 almost wiped out the mutation.

As far as we are aware this mutation is still in existence but cannot yet be considered established.

THE CINNAMON RANGE

We have identified four distinct cinnamon type mutations, plus a possible fifth which appeared during the 1990 breeding season in South Australia.

None of these mutations can be considered established and one, the first to appear, is probably extinct.

TYPICAL CINNAMON

This cinnamon mutation, which appeared synonymous with a typical sex-linked cinnamon with approximately 50% dilution, originated in the Newcastle region of New South Wales during the 1970's. Unfortunately its occurrence coincided with the development of the Yellow Turquoisine whose extreme beauty overshadowed this cinnamon mutation hence it did not receive the attention it deserved from aviculturists.

The mutation was still in existence during the early 1980's when the late Sid Gale saw these cinnamon Turquoisines in an aviary in the Hunter Valley region. He later acquired one of these cinnamons but had no success with it.

We have made extensive enquiries in the Hunter Valley district in an attempt to relocate this mutation — so far, without success. Possibly this mutation has died out.

RECESSIVE CINNAMON (Black-eyed) See Plate No. 3B

This cinnamon mutation, which has some dilution of the blue as well as of the green body colouring, the dark flight feathers reduced to fawnish brown, flesh coloured feet, diluted beak, and hatches with black eyes, was produced by Reg Collyer of Adelaide, South Australia, during the 1989 breeding season from high quality Red-fronted stock.

The cinnamon was one of six young males reared in the first nest of the season and of course the mutant being a male, indicates a recessive inheritance is involved.

Reg has since bred more of these black-eyed cinnamons as well as a similar mutant which hatched with deep red eyes (See plate No. 3C) that darkened by the time it fledged. Probably this mutant is yet another distinct cinnamon mutation — the fifth we have recorded.

Some prefer to call these type of mutations "Isabell" although we require further clarification of this loosely applied term before we will use it.

RECESSIVE CINNAMON (Plum-eyed) See Plate No. 3D

During the 1989 season Dave Farncombe of Gawler, South Australia had two chicks with plum coloured eyes appear in a clutch of young Turquoisines from normal parentage. These plum-eyed chicks, who were both males, retained the plum coloured eye after fledging but then the eye gradually darkened.

This cinnamon mutation, which is recessive as indicated by the first mutants being males, has more dilution of the green and blue than the previously discussed mutation and similar dilution of feet and beak. The feature that impressed Stan the most with this mutation was the softer powder blue colouring of the face and wings.

During the 1990 season one cinnamon male was mated to its sister and produced infertile eggs for the first clutch. The second nest of eggs was fertile and a clutch of split youngsters was reared which possibly indicates the hen used in this pair is not a split. The other mutant male was mated to his mother and three young were reared in the first nest, two of which were cinnamons. Both these pairs had further fertile clutches of eggs at the time of writing.

FALLOW See Plate No. 3E

The term "Fallow" is generally applied to cinnamon type mutations with slight dilution which retain a red eye after maturity. The degree of dilution and the depth of colour of the red eye varies from one Fallow mutation to another thus indicating the existence of numerous distinct Neophema mutations within the Fallow range.

Recently two Fallow Turquoisine hens were offered for sale by an Adelaide Bird Dealer. One of these birds was secured by Reg Collyer of Adelaide who is currently working on this mutation and has recently fledged one youngster from the mutant hen.

Most Fallow mutations are of recessive inheritance although distinct sex-linked Fallow mutations are possible.

GREY-GREEN (OLIVE) See Plate Nos 4A and 4B

Grey-green is the unfortunate and misleading term used in the Budgerigar Fancy to describe Olive mutations in which the blue colouring is replaced by grey in varying depths of colour, depending on the intensity of the blue being replaced.

This mutation is quite distinct from the true Olive mutation which retains all blue colouring.

We suggest that perhaps Grey-Olive is a more descriptive and suitable name for this mutation.

This Grey-Olive mutation is of a dominant inheritance with a visible single and double dark factor involved and is comparable to the Australian Olive mutation of the Peach-faced Lovebird.

During the early 1980's Brian Slater of Broken Hill, New South Wales, bred a Turquoise Parrakeet in which all green and blue areas were of a darker shade. At first he did not realise the significance of this darker bird but when more were bred from this normal pair Brian started to concentrate on these birds. Some were disposed of — one pair going to one of Brian's friends in Broken Hill.

In 1985 Brian bred a Grey-green (Olive) Turquoisine from a pair of the darker birds. A few weeks later Brian's friend also bred a Grey-green from his darker pair. Both pairs had produced a double dark factor youngster in their first nest of the season.

All the green areas in this mutation are replaced with olive green, the blue is replaced with charcoal grey while the yellow and red remains the same or slightly darker.

Brian states that the male single dark factor birds are easily recognised but female single dark factor birds can be difficult to identify.

This mutation can now be considered established in Australian aviaries.

PIED

A pied mutation has recently occurred in Australia which is said to be of recessive inheritance. This mutation is not as yet established and appears to bear no resemblance to the European Pied.

ACQUIRED YELLOW (PIED)

Recently, acquired yellow birds have occurred in this species. The extent of the yellow areas is often variable and usually increases with age. The acquired yellow colouring often eventually takes over all other colours with the occassional exception of red. The beak and feet loose all colour pigment, yet the eye never changes to red.

We have worked with this colour form in other genera for some years and have been unable to determine its inheritance or even if this colour form is transferred genetically.

At times the colour change is completed during the nestling period while at other times only a partial change takes place in the nest and sometimes the change commences during the juvenile moult.

Acquired colour changes are permanent and should not be confused with the yellow mottling associated with vitamin deficiencies, genetic deterioration and blood disorders.

RED-BELLIED OR RED-FRONTED See Plate No. 3F

This colour type is not a mutation but a "colour variant" which was produced through the exhaustive work of Australian aviculturists by continual selective line breeding of birds showing an orange or reddish patch on the belly.

In the days when wild caught Turquoise Parrakeets found their way into Bird Dealers' shops, odd birds, often hens, could be found with orange belly patches and from these birds the Red-fronted Turquoisines were developed. The orange-red patch, which varies in depth of colour, seldom extends beyond the belly on females but may extend from the vent to the throat on males.

It has been stated that this colour type is of recessive inheritance. This is not so. The colour or extension of the colour is brought about by mating two birds together which visibly carry the desired colour. Best results are not always gained by mating the most coloured birds together — often poor coloured individuals will produce good coloured young.

Australian Secondary Mutations

SINGLE AND DOUBLE DARK FACTOR DILUTE YELLOW See Plate Nos 4C and 4D

With the appearance of the Grey-green (Olive) mutation it could only be a matter of time before this single and double dark factor mutation was combined with the dilute Yellow mutation to produce a single and double dark factor Yellow Turquoisine.

Brian Slater of Broken Hill, New South Wales, bred the first SingleDark Factor Yellow Turquoisine from a pair of Single Dark Factor Split for Yellow Turquoisines in April 1989.

The Single Dark Factor Yellow mutation has a darker shade of yellow and noticeably darker blue areas than the normal Yellow Turquoisine.

Brian bred the first Double Dark Factor Yellow Turquoisine at the end of 1989 also from a pair of Single Dark Factors Split for Yellow.

The Double Dark Factor Yellow mutation is a mustard yellow bird with all blue areas replaced by a dark mauve colour.

Red remains constant in both these mutations.

The genetic inheritance of these mutations are a combination of a single and double dark factor dominant inheritance and a recessive inheritance. Hence to produce these colours the genetic potential for both the single or double dark factor dominant inheritance as well as the recessive inheritance must exist.

RED-BELLIED OR RED-FRONTED DILUTE YELLOW See Plate No. 4E

The Red-bellied or Red-fronted colour variation has been transferred into the Yellow mutation of the Turquoisine by selective line breeding just as it was developed in normal Turquoisines.

In females the orange-red seldom extends beyond the belly while in males the orange-red colour extends from the vent to the throat but as yet the colour is usually patchy on the chest, having not yet been developed to the extent seen on normal male Red-fronted Turquoisines.

Overseas Primary Mutations

DILUTE YELLOW

This mutation first appeared overseas in England about 1968, the result of an importation from Australia. It is now established throughout the world.

PAR BLUE (Partial Blue)

The Par Blue mutation of this species apparently occurred in Denmark about 1978 and is now slowly filtering into other parts of Europe.

The Par Blue Turquoisine is said to resemble the Green-backed (Seagreen) Scarlet-chested and is of recessive inheritance. This mutation shows a dilution of the blue on the head and the wing.

FALLOW

This variable mutation originated in England prior to 1979. It resembles a normal bird with slightly diluted yellow and blue colouring, a grey wash throughout the green, dark red eyes and reduced colour pigment in the feet.

At this stage the mutation is thought to be weak although inter-breeding with other mutations is expected to produce an improved Yellow with red eyes as well as a bluer bird with red eyes.

It is of recessive inheritance.

OLIVE

Little is known of this mutation except that its existence was reported in Denmark in 1983. Apparently colour photographs have been published and from these it is assumed to be a true Olive mutation and hence retain all blue colours.

We believe it would be of dominant inheritance with a single (dark green) and double dark (olive) factor involved.

Of course this mutation has the potential to darken dilute yellow and par blue birds.

PIED

This is a sex-linked mutation which is totally distinct from the Pied mutation in Australia and is thought by some to be allied to opaline.

The mutation was first reported in 1964 but was developed by Mr. B. Fregeres of Den Dolder, Holland from 1968 onwards.

Pied areas take the form of yellow spotting rather than smooth patches and is more prominent on the belly and chest with many yellow feathers being edged with green. In heavily marked birds, blue areas may carry white spotting, however heavy pied markings on the wings are quite rare.

Sexing of the mutation is difficult as the bright blue head and red wing patch of the male are reduced and for some inexplicable reason females carry a slight red wing patch, thus making both sexes similar.

RED-BELLIED OR RED-FRONTED

This colour form has evolved overseas through the selective line breeding of imported stock similarly to the way it was created in Australia.

RED-BELLIED OR RED-FRONTED DILUTE YELLOW

The transference of the orange-red belly and chest colouring of the Red-fronted Turquoisine to the Dilute Yellow mutation has been accomplished in Europe and England as it was in Australia.

Hybrids

Hybrids have been recorded between this species and the Scarlet-chested Parrakeet and the Elegant Parrakeet (*Records of Parrots Bred in Captivity* Prestwich, Additions 1954, Page 68).

During 1990 Bill Schwarzenberg of Victoria accidentally bred hybrids from this species and the Rock Parrakeet.

SCARLET-CHESTED PARRAKEET

Plate 5. Scarlet-chested Parrakeets (pair) male on right.

SCARLET-CHESTED PARRAKEET

(Neophema splendida)

Derivation:
Neophema:
  Neo — from *Neos,*
    Greek for new
  Phema — from *pheme,*
    Greek for voice
splendida — from *splendidus,*
    Latin for bright

Description:

Length 20 cm average
Weight 40 g average, females are usually
  lighter than the males
Appearance as in Plate No. 5

Classification

The Scarlet-chested Parrakeet, *Neophema splendida,* is the second "type" member of the Neophema genus, being a small brightly coloured, ground frequenting parrakeet with green body, blue face and wing markings and pronounced sexual dimorphism.

Although this species appears closely allied to the other typical genus member, the Turquoise Parrakeet, *Neophema pulchella,* these species are in fact more distantly related as indicated by the almost total infertility of the hybrids produced from these species. In addition DNA testing also indicates considerable differences between the species.

Gregory Mathews nominated the subspecies *Neophema splendida halli* from South Australia which had no blue on the back of the head and a less pronounced red breast. This subspecies is currently not recognised (*Australian Avian Record,* Volume 3; Page 57, 1916).

This species has a totally arid and semi-arid range across southern inland Australia.

Earliest Report

The Scarlet-chested Parrakeet was first described by John Gould in the *Proceedings of the Zoological Society,* London 1840, page 147 as *Euphema splendida.*

In 1891, Salvadori listed this species as *Neophema splendida* in the *Catalogue of Birds of the British Museum,* Volume 20, page 576.

The first illustration of this species is apparently that in Gould's *Birds of Australia* 1841, Volume 5, Plate 42.

Range, Habitat and Field Notes

Due to the remote arid regions inhabited by the Scarlet-chested Parrakeet little is known of its movements. It appears to be nomadic, being influenced by seasonal conditions and a preference for particular types of habitat.

Population eruptions in particular localities, usually at the extremities of the range, have often been observed and are obviously related to a highly successful breeding season followed by the congregation of the inflated numbers at a suitable feeding ground.

These fluctuations in the numbers of this species must occur throughout the extensive range, but, due to the remoteness, many such upsurges go unnoticed.

Obviously it is impossible to determine the exact range of a species which inhabits some of the most isolated country in Australia and whose movements must be so variable and influenced so much by the occurrence of rain in the arid interior.

Basically the Scarlet-chested Parrakeet ranges across the southern interior of Australia from the south-west corner of Queensland, through New South Wales, west of the Darling River to the north-western corner area of Victoria, then westward to the northern end of St. Vincent Gulf and up to Port Augusta in South Australia. From Port Augusta the recordings extend south-west down Eyre Peninsula to the Mt. George region then west across the Nullarbor Plains as far south as 30°S, down to the Kalgoorlie district of Western Australia, west to the vicinity of Lake Barlee and north-west to the upper reaches of Murchinson River.

The northern limit of the range appears to be around the Lake Macdonald-Kintone Range district near the Western Australian–Northern Territory border, the Macdonald Ranges and the Alice Springs region.

This species shows a preference for mallee, mulga and acacia scrubs, particularly where there is good ground cover of spinifex. Sightings have often been made in recently burnt out scrub country, the birds obviously being attracted by the regrowth or perhaps the seeds from the shrubs and trees Many Australian native plants have seed pods that split open during bushfires.

Scarlet-chested Parrakeets have been observed great distances from water hence it has been suggested that this species can obtain sufficient moisture from dew, frost and plants which store water to survive.

Stan recently examined succulent plants growing in the Great Sandy Desert. There is no surface water available for birds to drink, except that from frosts, heavy dew and occasional rain, so obviously the Scarlet-chested Parrakeet inhabiting this region must obtain moisture from another source. Most of the succulents investigated contained at least some moisture. One variety which had hundreds of small fleshy leaves, measuring at least 6 mm (¼ in.) long by 3 mm (⅛ in.) wide, on each small low plant, produced one large drop of clear water from each leaf when crushed between the fingers. This would be quite adequate to support the Scarlet-chested Parrakeet and other desert species.

On the other hand it has been recorded that Scarlet-chesteds, which were prevented from drinking at their usual watering place, became alarmed and took desperate measures to take a drink.

Feeding is carried out on or close to the ground and the diet consists of the seeds of grasses, herbaceous plants, acacias and other shrubs with a preference for spinifex seeds being observed. The seeding heads are reached by pulling the stem down with the foot.

Field sightings are usually of pairs (at breeding time) or of small groups, although large flocks of a hundred or more have been observed during periods of population escalations.

Bert Pollard of Barmera, South Australia, estimated between 800 and 1000 birds frequenting mallee country in the Upper Murray region of South Australia in flocks of from 10 to 80 birds. These sightings were made during the months of April, May, June and July from 1960 to 1965. The fearlessness and the variation in the extent of the red chest of the male, as well as the orange on the belly of the female in this species, was noted by Bob Pollard (*Australian Aviculture*, August 1966, pages 113–116).

They are considered a quiet, ground frequenting species which can easily be passed unnoticed but once sighted will usually allow close observation before moving away through the grass or fluttering off close to the ground.

Stan's only sighting of this species occurred during the afternoon of 19th June, 1990, in the Great Victoria Desert of South Australia.

A trip had been organised into this remote part of Australia for the express purpose of seeing in the wild two Neophema species, the Bourke's Parrakeet and the Scarlet-chested Parrakeet.

The party consisted of natural history bookseller Andrew Isles and Belinda Gillies-Isles, aviculturist Bill Schwarzenberg, four wheeler off road enthusiast Allan Isherwood and his son Chris and Jill and Stan Sindel.

It was generally considered essential to use two four-wheel drive vehicles on a trip such as this in case of mishap or breakdown. The desert is waterless and totally uninhabited and it is unlikely any other travellers will be encountered, thus all necessary goods and equipment must be carried such as tools, spare parts, C.B. radio, tents, sleeping gear, warm clothing (night temperatures drop well below freezing), food and water for eight days plus an additional 50% in case of emergency and fuel for 1100 kilometers.

All relevant permits must be obtained from the controlling authorities to allow entry to Aboriginal lands and Commonwealth Defence Department land.

This section of the trip commenced at Coober Pedy, South Australia, along a track with the misnomer Ann Beadell Highway, which extends westward for several hundred kilometers to the Western Australian border and beyond.

Stan and his friends followed this track to the Serpentine Lakes (salt lakes) on the Western Australian border through acacia scrub, mallee, mulga, red sand ridges and spinifex, all of which Stan describes as beautiful country.

On the fourth day in the desert, while returning to the camp after a day trip to Serpentine Lakes, Bill, who was in the lead vehicle, saw a small parrot fly from a small bush at the edge of the track. He followed it on foot to the area where he had seen it land and as he approached the bird flew past him from the top a shrub where it had perched. It was an adult male Scarlet-chested. It flew back across the track and landed on top of a dead stump 60 m away.

At this time the second vehicle arrived and all members of the party had clear views of this magnificent male. Stan photographed the bird from 60 m, then 50 m, but as he tried to move closer the bird flew back across the track, about 2 m above the ground and within 8 m of the group. It was in full late afternoon sunlight which highlighted the brilliant blues and scarlet chest of this adult male and provided a vision never to be forgotten.

Chris, who is a non-bird-person, commented he would never have believed that a group of adults would drive thousands of kilometres into remote desert for just a few minutes glimpse of a rare bird, and just how much pleasure and satisfaction they all gained from the sighting, had he not seen it for himself.

This sighting occurred at 4.30 in the afternoon 95 km west of Vokes Hill. After the initial sightings this bird could not be located again. The next morning the area was searched again but no Scarlet-chesteds could be found.

The flight of this species is of a fast, floating, fluttering and erratic nature. They usually prefer to fly through the scrubland rather than above it.

Breeding in the Wild

Breeding has been recorded from August to December and is obviously influenced by weather conditions. During good seasons double breeding may occur and account for sudden increases in numbers.

This species could be termed a loose colony breeder with pairs often nesting close to each other. Nests have been recorded as close as 4 m apart.

Nests sites in hollow limbs and holes in trees, often in eucalypts, have been recorded at heights of 2.5 m (8 ft) to 8 m (26 ft). The hollows selected are usually vertical and depths of about 40 cm (16 in.) have been observed.

Three to six, white, oval eggs are laid on a base of wood dirt. Green leaves are often found in the nest with incubating females and it is assumed that they are placed there in a similar manner as with the female Turquoise Parrakeet.

Sitting females have been observed to be fed by the male near the nest site.

The incubation period is given as approximately 18 days and the fledging period about 30 days. After leaving the nest the young, which resemble the adult female, remain with the parents for some weeks.

Red feathers appear on the chest of young males at about three months of age.

Aviculture

The first reference to this species in aviculture is of a lone female who was taken from a hollow in a mallee tree, where she was incubating eggs, by Mr William White of Adelaide, South Australia on the 29th September 1863. The nest site was discovered on Pudnookna Station (11 miles east of Morgan), River Murray, South Australia. Although always solitary and shy, this female lived for several years in Mr White's aviaries (A. G. Campbell, *Nest and Eggs of Australian Birds* Part 2 1901. Page 1082).

Scarlet-chested Parrakeets were considered a rare avicultural subject in Europe and England during the latter half of the 19th century. W. J. Greene in his *Parrots in Captivity*, 1884, laments the lack of importations of this species during that period.

The Zoological Society of London purchased a pair of Scarlet-chested Parrakeets in January 1871 from a bird dealer for £7-0-0 and a chick was hatched on July 21, 1872. There is no reference to this chick fledging in the Society's records although Karl Neunzig in his *Die Fremdlandishe Stubenvogel*, Volume 3, 1881, when citing this breeding of the Scarlet-chested states: "First arrival to the London Zoo in 1871, where it bred".

This breeding is refuted by Howard Jarman in The Scarlet-chested Parrot — *The Australian Bird Watcher*, Volume 3, No. 4, page 111, December 31, 1968. He states that the bird was in fact purchased from a dealer and cites Seth-Smith 1903 and 1932 as references. We can find nothing to support Jarman in Seth-Smith 1903 but in his article "The Splendid Grass-parrakeet" published in the *Avicultural Magazine*, Fourth Series, Volume 10, No. 4, page 73, 1932 he states that after research into the Zoological Society's records, he feels the supposed breeding in 1872 was a clerical error and in fact related to the purchase of another Scarlet-chested for 20 shillings. David Seth-Smith summed up by saying: "It is almost certain that the species has never been bred in captivity".

Apparently this species then vanished from aviculture until 1931 when several pair of wild birds were trapped near Oodnadatta in South Australia then taken to Adelaide and sold. The Governor of South Australia purchased one pair which were presented to King George V of England.

Simon Harvey, of Adelaide, South Australia, obtained two pair of these birds and bred this species in the same year (1931) for the first time in Australia and probably the world. He was awarded a bronze and silver medal for this achievement by the Avicultural Society of South Australia in 1932. During 1932, Simon Harvey reared in excess of 12 young Scarlet-chested.

Also in 1932 H. B. Scholz of Yaninee, South Australia, trapped 12 Scarlet-chested parrakeets about 290 km west of Port Augusta and south of the Gawler Ranges in South Australia. Presumably these birds were also introduced into aviculture.

In 1939 a large flock of Scarlet-chesteds was located near Wynbring, South Australia, the result of a population explosion. It was estimated that 500 of these birds were trapped for the world market ("The Scarlet-chested Parrot", Geoff W. Haywood, *Australian Aviculture*, July 1986, page 175).

From these three trappings this rare and sparsely populated parrakeet was established into aviculture and destined to become one of the world's most successful aviary psittacines.

In 1935 Edward Boosey of the Keston Bird Farm in England was awarded the Avicultural Society's medal for the first breeding of this species in Europe. The pair came from the Duke of Bedford, the male being from the original pair presented to King George V.

The Scarlet-chested Parrakeet has gone from strength to strength after these initial breeding successes to become totally domesticated throughout the world.

Stan's records reveal he first bred this species in 1960 from two, one-year-old pairs which had been purchased from a Sydney breeder who had bred them the previous season.

Both pairs were housed in fully roofed aviaries which were 1.8 m (6 ft) long, 90 cm (3 ft) wide and 2 m (6 ft 6 in.) high, with sand filled floors.

Hollow logs were provided for each pair which were approximately 30 cm (12 in.) long, had an internal diameter of 15 cm (6 in.) with an entrance hole near the top and fitted with a removeable lid. The logs were hung at about 30 degrees to the vertical and had a filling of coarse sawdust.

The two pair nested in the spring rearing four babies each. One of the pairs double brooded and fledged a second clutch of four.

Stan was delighted with the result but disappointment was just around the corner, as it so often happens in aviculture. The following autumn six of the 12 young died suddenly. In those days aviculturists had very little help, there being no avian veterinarians, no antibiotics available and only outdated literature. Treatment mainly consisted of the use of a hot-box and sulpha drugs or syrup of buckthorn — obviously cures were rare.

Even today Stan has troubles with young Scarlets during the autumn. As overnight temperatures start to drop, the juveniles suffer stress and become vulnerable to bacterial infections. Heat and antibiotics help to control this problem now although Stan believes that if young Scarlets are housed in protected aviaries or in a birdroom and provided with artificial heat, such as a small heating lamp, during the autumn and winter months the problems do not eventuate.

There is no doubt that some climates in Australia suit Scarlet-chesteds and others do not. Sydney's south-west is an area that does not, hence the need for care and caution with juveniles in this type of locality. At 12 months old this species stabilises and from then on is relatively hardy.

Generally speaking, Scarlets prefer dry climates hence breeders in inland areas are more successful with this species.

Scarlets have bred successfully in a variety of housing situations from small cabinets, suspended cages and aviaries to small aviaries, to large planted conventional aviaries.

We can find no records of really successful colonies as the males usually become quite aggressive at breeding time and disrupt the entire colony.

In our aviaries Scarlet-chesteds have been bred in flock situations when housed with Bourke's Parrakeets, Princess Parrots, Superb Parrots, Regent Parrots, King Parrots and Cockatiels as well as various finches and doves. Usually they do not co-exist well with other Neophemas due to fighting, except for Bourke's Parrakeets at odd times. When housed with larger parrots a constant vigilance should be maintained for any sign of aggression towards the smaller species.

A pair of Scarlets housed in a planted aviary will do little damage to established shrubs and be totally inoffensive to finches, softbills and doves.

The best long-term breeding results are gained from pairs held singularly in small sheltered aviaries.

Many aviculturists have experienced Scarlet-chesteds' liking for mealworms and termites but Peter Hobbs of Gloucester, New South Wales found another live food they were fond of.

Some years ago Peter fed a handful of milk thistles to his Scarlets which were covered in aphids. The Scarlets, who had never seen an aphid in their lives, flew down and greedily fed on them. From then on whenever Peter found aphids on his roses, citrus trees or whatever, he snipped off the infested branch and fed them to his Scarlets.

Pairs feeding babies ate aphids ravenously and fledged beautiful young when they were included in the rearing diet.

Sexing

Sexing of adult Scarlets is unmistakeable with the male carrying a bright scarlet-red chest, which is absent in the female, and an intensive, brilliant blue face which is reduced to mid-blue in the female. On very rare occasions a female may show an odd red feather on the chest.

Immatures are quite difficult to sex although some young males appear to have a more robust build and brighter blue face. Red feathers start to appear on the young male's chest at about three months old.

Hen Scarlets can be identified from hen Turquoise Parrakeets by the sky blue wing band on the wing coverts of the Scarlet as opposed to the mid-blue wing band of the hen Turquoise Parrakeet.

Display

The female is approached by the male who stretches to his full height, spreads the shoulders of the wings and prominently exhibits his red chest while bouncing along the perch rendering a low whistling twitter.

Nests

Nesting facilities of every conceivable type have been used successfully for this species. When given a preference Scarlets will often choose a much larger nest box or hollow log than would be expected. For instance, Scarlets housed with Princess Parrots in Stan's collection often choose a large vertical nest box or log about 90 cm (3 ft) long with an internal width of 25 cm (10 in.) provided for the Princess instead of the small boxes meant for them.

For convenience we use the inexpensive commercially available "african lovebird" type nest boxes, either the horizontal or vertical design, for this species and discard them after a season or two. Any type of box which is acceptable to the birds is fine, providing there is adequate access for inspection and cleaning.

Jack Stunnell of Sydney, who is a particularly successful breeder of Scarlets, uses a larger horizontal nest box with an entrance spout and the front half of the base of the box inclined at a 45° angle towards the entrance (see Figure 6F).

Vertical or partially inclined hollow logs about 30 cm (1 ft) long, an internal diameter of 15 cm (6 in.) to 23 cm (9 in.) with a removeable lid for access and the entrance hole on the side near the top are usually preferred.

Any of the recommended fillings are suitable for this species.

Nesting and Hatching

The Scarlet-chested Parrakeet's breeding season may at times extend throughout most of the year. Many breeders, particularly those using night lighting in their aviaries to promote breeding, have winter nests.

Under normal conditions breeding may commence as early as August and extend as late as April or May, with time out while moulting during mid to late summer.

In our less than suitable climate for this species we have had eggs laid as early as August 13th and as late as April 2nd, with the majority of eggs being laid during the six month period from September to February.

Interest in the nest site is shown by the male and female and soon afterwards the first egg is laid. Three to eight white, oval eggs have been recorded but four to six eggs form a normal clutch. The eggs are normally laid at two-day intervals with incubation, which is carried out by the female, usually commencing with the laying of the second or third egg. The most frequently recorded clutch in our aviaries was five eggs.

We have found the incubation period to vary more in this species than any other Neophema, with incubation periods of 17 to 22 days having been recorded under closely monitored conditions. Two complete clutches from different pairs hatched in 17-day periods both of which were recorded during our hottest months of January and February. The most commonly recorded periods were 18, 19 and 20 days.

The chicks hatch with silvery-white down, at a week old the eyes have started to open and early pin feather development of the tail and flight feathers are visible. At 12 days old pin feathers are well developed amongst the grey body down, and at 20 days the tail and flight feathers are 2 cm long. When 25 days old the chicks are almost fully feathered. We have recorded fledging periods of 26 to 32 days with the shorter periods usually being recorded during the hotter summer months.

When the young fledge they are similar to, but duller than, the adult female with large black-brown eyes and yellow beaks which change to brown by 10 weeks old. Young males start to moult in some red chest feathers at about three months old and by six months females have attained full adult plumage and males have moulted into sub-adult plumage. Males do not acquire the brighter, full adult plumage until their second year.

Sexual maturity has normally developed by 12 months of age.

Mutations

Australian Primary Mutations

PARTIAL BLUE (Par Blue)

PASTEL BLUE AND SEAGREEN (Green-backed) See Plate Nos 7A and 7B

The first partial blue Scarlet-chested Parrakeet was bred by Douglas ⁻kin of Heyfield, Victoria in 1958 from what appeared to be a normal pair. This recessive mutation was developed in Australia prior to being legally exported to England via Melbourne Zoo.

A Brooding hen with newly hatched chicks.

B Twelve-day-old nestlings.

C Nineteen-day-old nestlings.

D Twenty-six-day-old nestlings just prior to fledging.

E Red-bellied pair. The extent of red on the belly varies from a little (female) to a lot (male).

Plate 6. Scarlet-chested Parrakeet — nestlings and mutations.

A Pastel Blue (partial blue) pair.

B Green-backed Partial Blue pair.

C White-fronted Partial Blue pair.

D Cinnamon. This recessive mutant darkened in colour with age.

E Fallow hen showing slight dilution, pink feet and red eye.

Plate 7. Scarlet-chested Parrakeet — mutations.

Doug Ikin informs us that the first Blue Scarlet to be bred was a Pastel Blue hen and that as he progressed with the mutation both Pastel Blues and what we now call Seagreens or Green-backed Blues occurred in the same nests of young.

He also recalled that the young Pastel Blue cocks were weak in comparison to the young Seagreen cocks and it took some time to develop a robust strain of Pastel Blue cocks.

This information on Doug's early development of the Partial Blue Scarlets supports the theory that the Pastel Blue, Seagreen and perhaps the White-fronted Blue mutations are all just variations of the Partial Blue mutation.

Yet in contradiction to this theory, Doug also told us that the late Cook Hutchinson of Adelaide, South Australia, bred a partial blue mutation of the Scarlet which was similar in appearance yet genetically distinct from his, a few years after the initial breeding. Doug's comment was that when the two mutations were bred together it "buggered them up".

Our investigations into this second partial blue mutation suggest that perhaps Cook Hutchinson, who acted as an agent for Clary Burfield at this time, did not in fact breed this partial blue mutation but handled transactions on behalf of Clary Burfield.

Clary Burfield confirms he did develop a Partial Blue mutation of the Scarlet which was bred from his normal stock about this period.

Now so many years after the event we put the questions "Did this second Partial Blue mutation fully establish?" and if so "are our Partial Blues of today a mixture of both mutations?".

Some authorities argue that the Seagreen and Pastel Blue are two distinct mutations and quote as evidence records of normal green young being produced when the two are mated together. There is also records of normal young being produced from Seagreen × Seagreen as well as from Pastel Blue × Pastel Blue matings. Of course the vast majority of Partial Blue × Partial Blue matings produce only Partial Blues (either Pastel Blues, Seagreens or both).

Now to confuse the issue even further, intermediate colour strains have been identified, such as Cream-bellied Pastel Blue and Ivory-bellied Pastel Blues.

Many aviculturists are now specialising in Blue Scarlets, so hopefully breeding records from these specialists will eventually provide some answers. In the meantime we suggest that each established colour variation be kept as distinct from the others as possible.

In summing up and assessing all the evidence available, we feel the majority of Partial Blue Scarlets are of the same mutation, which involves two basic colour types, namely, the Pastel Blue and the Seagreen plus a few other colour variations. But, the unpredictable and supposedly impossible results experienced by some breeders from certain pairs suggest the involvement of a second distinct mutation in some strains of Partial Blue Scarlets.

WHITE-FRONTED BLUE See Plate No. 7C

This mutation first appeared in John Lewitzka's aviaries in Adelaide, South Australia, during October 1981. It was bred from two green birds, thought to be split "blues" which were from John's deceased father, Fred Lewitzka's collection.

In 1985 Jack Stunnell of Sydney, New South Wales, bred two young of the same mutation from a Pastel Blue hen and a Seagreen cock, who on their second nest produced two normal green young, one White-fronted blue and one Seagreen youngster, which suggests the parents are both split for the new White-fronted Blue mutation. The two normal green young from this nest were mated to blue birds but failed to produce even one blue youngster over several nests, indicating they were not even split for blue.

Although Jack held split birds from John Lewitzka's stock in his aviaries, the White-fronted Blues were bred from his own independent stock.

Our first assessment of this new colour form was a distinct true "blue" mutation, devoid of all yellow colour pigment. These thoughts were supported by Des Cartwright of Sydney whose breeding results of White-fronted Blue to White-fronted Blue matings produced only White-fronted Blue young.

During the 1989 season Des again produced a number of White-fronted Blue young from the same pairs as the previous year, all of which were White-fronted Blues and appeared identical to each other and to the previous year's young. Yet some of the young males from the 1989 breeding moulted in pale apricot chests although totally blue and white in every other respect.

This indicates to us that the White-fronted Blue mutation of the Scarlet is in fact an extreme refinement of the Partial Blue mutation and not a true blue mutation that is devoid of all yellow pigment.

Another interesting theory put forward by Brian Slater of Broken Hill, who is a highly experienced breeder of Neophema mutations, is that the White-fronted Blue is the product of the progeny from a Seagreen mated to a Pastel Blue. Brian hopes to prove or disprove his theory this season.

Stan's comment on this theory is "if it is that simple to produce White-fronted Blues, why did it not happen much sooner?".

John Lewitzka and John Stracken of Adelaide, who believe two mutations are involved in the South Australian Partial Blue stock, have test bred White-fronted Blue with Partial Blues and provide us with the following results:

Seagreen split for White-fronted Blue mated to a Pastel Blue split for White-fronted Blue produced White-fronted Blues, Pastel Blues split for White-fronted Blue, Normal greens split for Pastel Blue and White-fronted Blue, Seagreen split for White-fronted Blue.

Seagreen split for White-fronted Blue mated to White-fronted Blue produced White-fronted Blues and Seagreens split for White-fronted Blue.

VIOLET BLUE

By far the most striking mutation of the Scarlet-chested Parrakeet to have occurred is the Violet Blue which is the violet form of the White-fronted Blue.

Whilst Stan was in attendance at the Adelaide (South Australia) Avicultural Convention of 1989, he and Jack Stunnell were taken to visit John Lewitzka's aviaries. There they saw the beautiful Violet Blue. The bird was bred from a pair of White-fronted Blues during 1988 in John's aviaries. Unfortunately this stunning mutation has not established as yet.

OLIVE See Plate No. 8A

This mutation was developed by Graeme Hyde of Victoria from a darker coloured male he obtained in 1978. Graeme bred from this bird and developed a colour strain he originally named "khaki".

In 1982, after a discussion with Barry Hutchins of South Australia, Graeme became convinced his "khakis" were a true mutation, so he renamed them "Light Olives".

From the Light Olive, which retain some green on the rump, the Olive mutation has been developed in which all green is replaced by olive green and the sky blue wing band becomes a pale bluish grey while all other blue colouring is retained.

A Olive pair: Note retains blue.

B Olive-blue (olive version of pastel blue) pair.

C Orange-fronted (Orange-chested).

D Pied-winged White-fronted.

E Acquired Blue Pied.

Plate 8. Scarlet-chested Parrakeet — mutations.

100

The Olive Scarlet-chested is a true mutation of dominant inheritance with three distinct colour phases, light, mid and dark, similar to the three distinct depths of colour type seen in the Grey-green Budgerigar. A single (light and mid) and double dark factor is involved in this inheritance.

Stan has found that with successive olive to olive matings the olive colour deepens as does the blue of the face and wing border. The sky blue wing coverts become grey with a mottled overlay of yellow in good dark specimens.

LUTINO

The sighting of a lutino Scarlet-chested Parrakeet was reported by the prominent aviculturist Syd Smith of Victoria to an Australian Avicultural Society meeting in Melbourne.

The mutant was sighted in Tommy Glascoe's aviaries in Adelaide, South Australia during the late 1960's. It was a female and reported to be yellow with a white wing band and face. Unfortunately this mutation was not established in Australia. (*Australian Aviculture,* August 1974, page 124).

CINNAMON GROUP

CINNAMON — RED EYED See Plate No. 7D

A cinnamon Scarlet male, which retained a deep red eye was exhibited in a bird exhibition at the Avicultural Convention in Adelaide, South Australia, April 1989. This bird was bred from a normal pair, hence, because the mutant is a male, a recessive inheritance is indicated. Cinnamons which retain a red eye after fledging are often of a recessive inheritance.

Unfortunately this bird, which was bred in 1988, has darkened with age but can still be identified as a red-eyed mutant of the cinnamon group. This mutation cannot be considered established as yet.

CINNAMON — BLACK EYED

A Fawn mutation of the Scarlet-chested (almost certainly a black-eyed recessive cinnamon mutation) was reported to an Australian Avicultural Society meeting in Melbourne, Victoria, many years ago by the late Eric Baxter. (Personal Communication from Graeme Hyde, Editor). This mutation was not established.

FALLOW (Isabel) See Plate No. 7E

A slightly diluted hen Scarlet-chested with deep red eyes and pink feet was reared in the Hunter Valley, New South Wales during the late 1980's, from a normal pair. It is a mutation of the cinnamon group as indicated by the red eyes and diluted feet and is similar to a mutation termed Isabel in Europe.

There is much confusion as to what constitutes an Isabel mutation, with European authorities offering varied opinions. Until clarification of this term is available with identification details and standardisation of unusual mutant names at a world level, we feel that rather than add to the confusion this mutant is best identified as a type of Fallow.

Work is being carried out on this mutation by an experienced Sydney breeder but as far as we are aware the genetic inheritance is yet to be determined.

PIEDS See Plate No. 8E

We have seen many pied-like Scarlet-chesteds over the years, most of which were the result of vitamin deficiency or general deterioration of health.

Acquired coloured birds, where yellow feathers (on normal birds) gradually moult in and sometimes eventually take over almost all the normal colouring, occur periodically. Some acquired coloured birds attain the pied feathers during the nestling period and actually fledge as pied birds.

We are not aware of the establishment of any of these type of pieds in Australian aviaries or if they are in fact a genetic inheritance.

HARLEQUIN PIED

In the New South Wales country town of Singleton, two "Harlequin Pied" type mutation (or perhaps the equivalent of the Dutch or Danish Pieds in Budgerigars) have occurred in separate collections during recent years.

The pied colouring in this mutation occurs predominately on the body with little or no pied on the wings. The Harlequin Pied and Danish Pied mutations are of recessive inheritance in Budgerigars whereas the Dutch Pied is of dominant inheritance.

This pied type mutation first appeared in a green bird during the mid 1980's, the second mutant occurred during the late 1980's in a Partial Blue Scarlet, apparently from unrelated stock.

102

As yet neither strain of this mutation has been established.

PIED-WINGED WHITE-FRONTED BLUE See Plate No. 8D

Three White-fronted Scarlets with pied wings were reared in two nests from the same pair by Kim and Roger Simmons of New South Wales during the 1990 season.

The pied areas occur symmetrically on the primary and secondary flight feathers of both wings with each feather retaining a normal coloured tip for approximately 25% of its length.

Stan's first impression of this mutation was of an extension of the white wing bar present in juveniles, in which case it would be expected the pied to moult out of young males as they come into adult plumage. This was not the case; the first two to fledge were a pair and both retained the identical amount of pied in each wing.

The female carries more of the symmetrical pied markings on the wings than the male; a feature which may yet prove to be related to an extension of the wing bar rather than any of the variable pied mutations.

As work continues with this mutation the possibility of the extension of the pied area together with the genetic inheritance will be established.

ORANGE-FRONTED See Plate No. 8C

In this mutation, as the name implies, the scarlet chest of the male is replaced with orange, the green body colour shows slight dilution and the yellow under-body colour is diluted to lemon yellow. Females of this mutation are not readily identified but can usually be distinguished by the lemon-yellow belly colour.

This recessive mutation was first bred by Brian Anderson of Singleton, New South Wales in 1982 and although not yet fully established is currently held in several breeder's aviaries.

We believe this mutation could be used to great advantage with the Partial Blue series of mutations to whiten the chest colour.

ACQUIRED YELLOW

A number of apparently acquired yellow birds have occurred over the years in which more and more yellow feathers are moulted in over successive years but usually some normal coloured feathers are retained.

The late Fred Lewitzka of Adelaide held a fine example of this colour form in his aviaries just prior to his death in 1979. John Lewitzka informs us the bird, which was a male, was almost totally yellow except for some normal coloured flight feathers and a few blue feathers on the face.

Many of the so-called Pied mutations encountered in this species are in fact birds with acquired yellow feathering in varying amounts. We are not familiar with any authentic pied mutations established in this country.

As yet we have been unable to determine with any degree of certainty if acquired colouring is of genetic inheritance or not. Stan's experience with this type of colour form in lorikeets leads him to believe it is quite unlikely to be a genetic inheritance.

Barry Hutchins of South Australia is in possession of a mounted specimen of an acquired pied which originated in Tommy Glascoe's aviaries in Adelaide, South Australia, many years ago. The bird was normal until one year old when the pied colour started to moult in. Eventually it was 95% yellow, but always retained a few normal feathers, then at 13 years old it went blind and died soon after. Only normal coloured young were ever produced from this bird.

RED-BELLIED See Plate No. 6E

This colour form is not a mutation, in the true sense of the word, but the extension of a colour variation which is present in this species (even in wild specimens), that has been brought about and accentuated through selective line breeding.

By constantly selecting breeding stock in which the male's scarlet patch on the chest extended further towards the belly and female's showing more extensive orange colouring on the belly, aviculturists have produced and fixed the Red-bellied colour variant.

The Red-bellied colour variety is not of a genetic inheritance and is reproduced by mating the best to the best, although often individuals showing lesser amounts of red on the belly produced exceptionally good Red-bellied young.

Barry R. Hutchins of Adelaide, South Australia, provides the following information on the results of selective line breeding of Scarlet-chested Parrakeets over a 16 year period.

The experiment commenced with the acquisition of four pairs of fine normal Scarlets. The aim was to selectively breed this stock for the improvement of size, brilliance of colour and the extension of the red chest of the male and the orange abdomen of the female. From the first season the offspring were culled ruthlessly, with only birds carrying the desired colour and size tendancies being retained for breeding stock.

Patience was an important quality in Barry's character for after the first breeding season there followed a five year period when the project seemed almost stationary. The breeding results were satisfactory but improvement of the selected features were only slight.

A lesser aviculturist would have quit the project but Barry persisted through these bad years then suddenly it all started to happen. The seventh season saw marked improvements in the characteristics being selectively bred for.

Sixteen years after the commencement of the project a strain of Scarlet-chesteds had been developed which was much larger than the original birds, the males carried a rich red from throat to vent, the greens and blues were of intense colour, a red wing-band had appeared, only on the males, similar to although not as extensive or rich as that of a male Turquoise Parrakeet, and an olive collar and nape had developed.

The females had red bellies, adults carried a red patch on the chest, the greens and blues were also intense in colour and the olive collar had appeared.

All this was achieved from the original birds which were typical normal Scarlet-chesteds, and without fresh bloodlines being introduced during the 16 year period.

Overseas Primary Mutations

PARTIAL BLUE

The Partial Blue mutations have been developed in England and Europe from Partial Blue stock exported via Melbourne Zoo (Australia) to London Zoo during the 1960's. Apparently one pair from the first importation was retained by London Zoo and the remaining birds went to the aviaries of Prof. J. R. Hodges (*Grass Parrakeets*, H. P. M. Zomer 1987).

Pastel Blue, Seagreen and White-bellied Blue strains have been developed from this imported stock in England and Europe just as they have been in Australia.

LUTINO

A Lutino Scarlet-chested appeared in the U.S.A. during the 1980's and currently the mutation is in the process of being developed.

The Lutino Scarlet-chested is also available in Europe.

THE CINNAMON GROUP

SEX-LINKED CINNAMON

This mutation was first bred in Germany in 1968 and was developed there by Mr. Budnik and Mr. Janssen in conjunction with the Isabel mutation which apparently appeared simultaneously.

This variable cinnamon mutation is sex-linked, the chicks hatch with red eyes which change to normal colour at eight to 10 days old. The feet are flesh coloured and the claws and beak are diluted.

ISABEL

This mutation is of the cinnamon group and was apparently inter-woven with the previous cinnamon mutation and emerged as a by-product during its development.

The Isabel mutation of the Scarlet-chested adds further to the confusion regarding identification of mutations bestowed with this European originated name.

Isabel Scarlet-chesteds may be described as being slightly diluted from normal with dull flesh coloured feet, diluted claws and beak and normal coloured eyes, while the red chest of the male remains normal. It is a sex-linked mutation.

FALLOW

This mutation is similar to a pale coloured normal with diluted feet and red eyes.

PIED

This unusual pied mutation is of a dominant inheritance with single and double factors involved. Young mutants leave the nest normal in colour then moult in the pied areas on their first, second, third or even fourth moult and the extent of the pied colouring may continue to increase during subsequent moults.

Apparently this mutation first occurred in Europe in 1967.

RED-BELLIED

The Red-bellied Scarlet-chested was developed in Europe from the selective line-breeding of birds imported from Australia, which carried excessive red on the chest or abdomen.

Overseas Secondary Mutations

GOLDEN YELLOW

This secondary colour was produced by combining the Cinnamon mutation with the Isabel mutation. In other words it is the cinnamon version of the Isabel mutation and in appearance is a yellow Cinnamon with a brown glow, which surprisingly retains red eyes after fledging. (Both this Cinnamon and the Isabel mutation fledge with normal coloured eyes).

SEAGREEN ISABEL, PASTEL BLUE ISABEL AND SKY BLUE

By combining the Partial Blue range of mutations with the Isabel mutation various partial blue versions of the Isabel have been produced.

The Seagreen Isabel is basically a diluted form of the Seagreen with lighter coloured feet and beak.

The Pastel Blue Isabel is a diluted form of the Pastel Blue with lighter coloured feet and beak.

The Sky Blue is a particularly beautiful secondary mutation which is the White-fronted Blue form of the Isabel mutation and can be described as a sky blue version of the White-bellied Blue, retaining the white front and belly and a diluted, yet dark blue face.

SEAGREEN CINNAMON, PASTEL BLUE CINNAMON AND SILVER

These secondary mutations are the cinnamon forms of the partial blue range of mutations.

Seagreen Cinnamon is the cinnamon version of the Seagreen and retains the blue facial and wing colours as well as the salmon chest of the male. The Seagreen colour of the body is replaced with cinnamon-olive while the diluted feet and beak remain.

Pastel Blue Cinnamon is the cinnamon version of the Pastel Blue which also retains the blue facial and wing colouring and the salmon chest of the male, while the blue-green body colour is replaced with a cinnamon-mauve. The diluted feet and beak remain.

The Silver Scarlet-chested is the cinnamon version of the White-fronted Blue in which the blue face and wing band is retained in a diluted form, the white front and white belly remains while the remaining blue body colour is replaced with a bluish-grey. The normal eye and diluted feet and beak are retained.

WHITE

This secondary mutation is the Lutino version of the White-fronted Blue. It is not a true albino because the White-fronted Blue is not a true blue mutation.

Hybrids

Hybrids have been produced between this species and the Turquoise Parrakeet and the Elegant Parrakeet.

ELEGANT PARRAKEET

Plate 9. Elegant Parrakeets (pair) male on right.

ELEGANT PARRAKEET

(Neophema elegans)

Derivation:
Neophema:
   Neo — from *Neos,*
     Greek for new
   Phema — from *pheme,*
     Greek for voice
elegans — Latin for handsome

Description:

   Length 22 cm
   Weight 46 g average
   Appearance as in Plate No. 9

Classification

The Elegant Parrakeet, *Neophema elegans,* is rather typical of the four species of blue frontal banded, green bodied group of Neophema (Neonanodes) who inhabit much of southern Australia. This parrakeet is the least specialised of the group.

This species is appropriately named, the word elegant describing perfectly the more slender shape and the subdued yet beautiful colouring, particularly as seen in the adult male. The brighter yellow-olive and yellow plumage together with the distinctive two toned blue frontal band in which the upper metallic light blue band extends beyond the eye, distinguishes it from the rest of the group.

There are distinctly isolated eastern and western populations in which only minor variations have occurred over a long period of time. However, Gregory Mathews in his *A List of the Birds of Australia* page 138, 1913 separated the duller western population as the subspecies *N.e. carteri* which was later discounted.

Earliest Report

After being confused with the Blue-winged Parrakeet, *N. chryso-stomus* for many years, Gould described the Elegant Parrakeet, *N. elegans* in the *Proceedings of the Zoological Society,* Part 5 1837 page 25, as *Nanodes elegans,* with the incorrect location of origin Tasmania, which he subsequently corrected at a later date.

The first illustration of this species (head only) appears to be that in Gould's *Synopsis of the Birds of Australia* Part 2, January 1837 as *Nanodes elegans*. Apparently the first full illustration is that in Gould's *Birds of Australia* 1848, Volume 5, as *Euphema elegans*.

The current name of *Neophema elegans* was first used by Salvadori in *The Catalogue of Birds in the British Museum*. Volume 20, page 572, 1891.

Range, Habitat and Field Notes

The Elegant Parrakeet has two distinct populations which are separated by the extensive natural barrier of the Nullarbor Plains, which effectively eliminates all interaction between the two groups.

The western population which is constantly extending its range, probably because of the increased grassland created by the clearing of forest and scrubland, is found in the south-west of Western Australia. This population has been recorded as far north as the Onslow and Robe River districts and as far east as Rocky Point near Israelite Bay. If a slightly concave line is drawn from this northern limit to the eastern extremity of this range, all of south-western Australia on the western side of this line is the known range of the western population of the Elegant Parrakeet.

In contradiction to this accepted range the RAOU's *Atlas of Australian Birds,* which we feel provides the most recent and concise information available on the range of Australian birds, indicates this population extends only as far north as the latitude of 30°S. All authorities agree on the eastern limits of this western population.

The eastern population of this parrakeet extends west to the Streaky Bay district of South Australia, north into the Flinders Ranges, then east to the Broken Hill-Menindee Lakes region of New South Wales, the Mildura district of Victoria and occasionally in the vicinity of the western shores of Port Phillip Bay. Within these western and eastern limits the range extends southward to the coastline and to Kangaroo Island where it is a seasonal visitor from late spring to autumn. Records suggest the Elegant Parrakeet is only an occassional visitor to the north-eastern and eastern sections of its range.

Elegant Parrakeets inhabit tree studded grasslands, cleared areas in open forest, mallee, scrubland, coastal sand dunes, arid saltbush areas and farmland.

The diet consists of the seeds of native and introduced grasses, herbaceous plants and shrubs, vegetable matter and small fruits and berries. All feeding takes place on or close to the ground.

This parrakeet is often flushed from roadsides or while feeding on the ground, when it rises quickly uttering a short single syllable, soft whistling yet penetrating alarm call, which is repeated twice in quick succession, then after a short pause again and again.

The Elegant Parrakeet is usually nomadic, with movements being dictated by weather conditions and hence the availability of food. Regular seasonal movements have been recorded in some areas, particularly at the extremities of their range.

In Western Australia this parrakeet is more numerous than in the eastern states except for the odd local concentration. Elegants from the eastern population are sometimes observed feeding in the company of Blue-winged Parrakeets, which they closely resemble in both appearance and general habits.

Stan's notes of a trip to Western Australia during 1971 state that his first sighting of Elegant Parrakeets in the field occurred on the 31st of August near Lake Grace, when a small flock flew from the side of the road, where they had been feeding on grass seeds. They landed in a large eucalypt where they were observed for sometime. Suddenly one bird left the flock, flying with rapid wing beats to a great height whilst calling with a feeble whistling note then levelling out and flying off into the distance. Stan has observed Blue-wings leave their feeding grounds in Tasmania in an identical manner.

During the same afternoon Stan saw another pair near Ongerup and yet another pair near Albany.

Then on the 3rd of September 1971 when travelling south of Perenjori near the village of Caron, Stan observed a flock of over 20 Elegants picking grit or perhaps foraging for seeds beside a railway line. They flitted about from place to place then ran along the ground in a typical grass parrakeet manner, quite oblivious of their observer.

We have never observed members of the eastern population in the field.

When travelling a distance the Elegant's flight is fast and direct with rapid wing beats and often at considerable height. Short distances are usually covered with an erratic floating flight which terminates with the tail spread when landing.

Breeding in the Wild

Elegant Parrakeets nest in hollow limbs or holes in trees with a preference for high dead horizontal branches in isolated trees or small stands of trees in clearings or in open country.

A clutch of four to five white elliptical eggs are laid on a base of rotten wood dirt at a depth of up to 60 cm (2 ft) from the entrance hole. Breeding usually takes place between August and December and double brooding may occur during good seasons. Incubation periods of 18 and 19 days have been recorded which are not consistent with our precise aviary records. Youngsters usually fledge about 30 days after hatching and moult into adult plummage at four months of age.

Small flocks are usually formed after breeding, consisting of adult and juvenile birds which forage for food nomadically.

Aviculture

Although a specific world first breeding is not recorded avicultural writers of the latter decades of the 19th century refer to numerous breedings in Europe and England during and prior to the 1880's. For instance Dr Russ in his Handbook states that the Elegant "has bred in several instances". W. J. Greene in his *Parrots in Captivity* 1884 Volume 1 page 85 refers to Mr Gedney's breeding in England which was obviously prior to 1884. Hopkinson in his *Records of Birds Bred in Captivity* 1926 refers to "Neunzig, page 751" where he quotes from Russ "they have been bred in Holland, Belgium and England."

The first recorded Australian breeding of the Elegant Parrakeet was by Mr S. Harvey of South Australia in 1930 who was awarded a bronze medal by the Avicultural Society of South Australia for his achievement.

Other early recorded breedings are; Australia, Lou Koenig since about 1939; Great Britain, Edward Boosey Keston Bird Farm 1936; USA, David West 1950 reports three reared. (Pretwich *Records of Parrots Bred in Captivity* 1954.)

Our first breeding was in Stan's aviaries during 1961 when a pair housed in a small fully roofed breeding aviary 1.8 m (6 ft) long, 0.9 m (3 ft) wide and 2 m (6 ft 6 in.) high with a concrete floor, flew three young. Stan's records also show that during 1965 and 1966, 16 young Elegants were reared in each season. The original pair he had first bred from which were still housed in the same small aviary flew two broods of four young each year. Another pair housed in a 12.3 m (40 ft) by 4 m (13 ft) planted aviary also flew two nest of four young each year.

During the late 1960's Stan had a pair of young from these earlier breedings housed in an aviary 3.6 m (12 ft) long, 0.9 m (3 ft) wide and 2 m (6 ft 6 in.) high, which also double brooded for several consecutive seasons and flew two nests of four young in most years. All of which indicates just how prolific, versatile and indifferent to housing the Elegant Parrakeet is.

Although Stan has held and bred Elegants in his aviaries every season since those early breeding years it was not until 1986 that five young were fledged from one nesting, yet five egg clutches had been recorded on several occasions over the years.

Our attempts to colony breed Elegants have always failed. This species becomes very aggressive towards its own kind as the breeding season approaches which usually results in so much interference that none of the colony breed or all but the dominant pair has to be removed from the aviary.

Yet when Stan visited Western Australia recently he was shown a number of breeding colonies of Cinnamon and Pied Elegants. All the colonies were housed in large aviaries and limited to two or three pairs, sometimes with a surplus of hens.

Stan has bred Elegants in flock conditions housed in breeding aviaries 3.6 m (12 ft) to 5.4 m (18 ft) long with either King Parrots, Superb Parrots or Princess Parrots. They have also bred in large planted aviaries which housed mixed collections of Softbills, finches, other species of Neophema, Regent Parrots, Crimson Wings and Cockatiels as well as the previously mentioned larger Australian Parrots.

Sexing

Adult males are brighter in general colouring than the female, particularly the blue frontal band and yellow lores. The males have more pronounced and brighter yellow underparts with the majority of adult males carrying an orange spot in the centre of the abdomen. This spot is seldom if ever seen in adult females.

Immatures are similar to adult females with large black eyes and a barely visible blue frontal band. Beaks are yellow when the young fledge and gradually change to brown about six weeks after fledging, then to black a few weeks later. Young Elegants are very difficult to visually sex.

Display

The male approaches the female in an animated manner stretches to his full height, drops his wings slightly, partially fans his tail, then shakes his head in mock feeding. Courtship display often ends with the male feeding the female.

Nests

This species is not difficult to please with nesting facilities. We have successfully used large and small hollow logs, hung vertically, horizontally and partially inclined. The ideal hollow log for this species would be approximately 30 cm (1 ft) long, with an internal diameter of 15 cm (6 in.) and with an entrance hole in the side, near the top. The log can be hung either vertically or partially inclined.

Nest boxes of every conceivable type have been used successfully for this parrakeet. Entrance spouts to the boxes will be accepted although they are unnecessary. These days we opt for simplicity and often use commercially marketed horizontal boxes designed for African Lovebirds and hang them at a slight angle to keep the eggs together at the rear of the box (see Figure 6A).

Whatever type of box or hollow log used it should always be fixed under shelter and easy access for cleaning and observation is essential. Any of the recommended nest fillings are suitable for this species.

Nesting and Hatching

Interest shown in the chosen nest site is a prelude to breeding, with the female spending more and more time in the nest until the first egg is eventually laid.

We have recorded eggs being laid as early as the 29th of July and as late as the 2nd of December, in the extreme climate of the south-western suburbs of Sydney. Usually four and sometimes five white elliptical eggs are laid normally at two-day and occasionally three-day intervals. We have recorded incubation periods of 19 days, 20 days and 21 days — never the often quoted 18-day incubation period.

Chicks hatch with silvery white down which is dense on the upper body and sparse on the back of the head, neck and under parts. The eyes usually commence to open at about seven days, flight, tail and early pin feather development is visible at 10 days, while at 14 days of age advanced pin feathering is apparent throughout the now grey coloured body down. When four weeks of age the chicks are fully feathered.

We have recorded fledging periods of 33 to 38 days. Fledglings leave the nest similar to, but duller than the adult female with only the slightest trace of a blue frontal band, large black eyes which lack iris rings and with a yellow beak.

116

At 10 weeks of age the beak has turned brown and the blue frontal band is quite noticeable. Youngsters moult into adult plummage by six months of age.

Mutations

Australian Mutations

CINNAMON See Plate No. 11A

The cinnamon mutation of the Elegant Parrakeet was first bred by Mr Herb Forrest of Adelaide, South Australia, during the 1981–82 breeding season, when a normal coloured pair produced a cinnamon hen in their clutch. The following season two cinnamon hens were produced and the same result the year after.

This is a typical although not well advanced, sex linked recessive cinnamon mutation in which the chicks hatch with plum red eyes and cream coloured down. By the time the young cinnamons fledge the eye colour has changed to dark brown but will still appear reddish when viewed under a strong light. Like most cinnamon mutations they are at their most attractive when they first leave the nest.

Cinnamon Elegants are now well established in Australian aviaries.

WESTERN AUSTRALIAN CINNAMON

During 1984 Hank Jonker of Forrestdale, Western Australia, bred and then later developed a cinnamon Elegant from stock of the western race. In appearance it is very similar if not identical to the cinnamon mutation developed from the eastern population.

There appears to be at least one important difference in these distinct sex linked cinnamon mutations. Breeders from Western Australia assure us that their cinnamons hatch with dark if not totally normal coloured eyes, not the plum coloured eyes of the eastern cinnamons.

Hence this must be a distinct cinnamon mutation and therefore it is important to keep it totally separate from the eastern cinnamon.

Bob Phillpot of Perth, Western Australia, provides us with his observations of this mutation. Firstly Bob has also noted the dark eyes in the chicks of this mutation as well as the orange coloured feet and beak. Young cinnamons bred in Bob's aviaries from split cock to normal hen matings fledge with mauve blue coloured frontal bands while those bred from split cock to cinnamon hen matings fledge with white frontal bands. All young revert to the normal blue frontal band after the juvenile moult.

A Brooding hen with one, three and five-day-old chicks.

B Fourteen-day-old nestlings.

C Twenty-one-day-old nestlings.

D Thirty-day-old nestlings just prior to fledging.

Plate 10. Elegant Parrakeet — nestlings.

Cinnamon hens which develop grey coloured eyes have also occurred in Bob's stock, but he is not certain at what age the eye colour changes.

CINNAMON YELLOW See Plate No. 11B

During the early 1960's a yellow coloured Elegant Parrakeet was observed in a wild flock and eventually obtained by Max Needorfer of Port Pirie, South Australia. Max described the mutant as a dark eyed, yellow bird with diluted blue marking and flesh coloured legs and feet. The blue areas on females were washed out whereas they were bright blue on the males.

Over a 13 year period Max bred many mutants and found this cinnamon yellow mutation to be of a sex linked inheritance. Unfortunately in the end this mutation was lost.

In the mid 1980's yet another wild yellow mutant was seen in the Port Pirie district, so obviously the genetic potential for this mutation still exists in the region.

LUTINO

A Lutino mutation was reported in a New South Wales aviary many years ago but apparently this mutation was not established.

There have been recent reports of the occurrence of a Lutino Elegant in Western Australia. Our attempts to confirm this report led us to Bob Phillpot of Perth, Western Australia, who, in 1989, hatched a chick with blood red eyes from a Western Australian cinnamon hen and a split cock mating. This chick and its nest mates were lost at the early pin feather stage when it was a mass of yellow pins. Bob does not state that this chick was a Lutino but it did have blood red eyes and yellow feathers.

PIED See Plate No. 11C

This recessive mutation was first bred by Doug Anderson of Bickly, Western Australia, in 1985 when two pieds were produced from a colony of five pairs of normals.

During 1986 two more pieds were bred from Doug's stock which had been given to South Perth Zoo. Both these birds as well as one of the previous year's pieds died.

In 1987 Doug lent the remaining Pied bird to Hank Jonker of Forrestdale, Western Australia, who paired it with a cinnamon hen. Seven normal young (the males split for cinnamon) were reared and eight more in 1988.

A Cinnamon pair. Male on right.

B Cinnamon Yellow. Mounted specimen.

C P ed with Normal.

D Cinnamon Pied.

Plate 1. Elegant Parrakeet — mutations.

The young from the first season, which were three males and three females were placed in a colony to breed and duly produced five cinnamon hens, five normals and two normal pieds.

In 1989 the original pied male was placed with two of the cinnamon hens bred in the 1988 season. Eleven normals which would all be split for pied and the males split for cinnamon were fledged.

During the 1990 season a normal pied male mated to two split pied females eventually fledged a normal pied after producing 20 normal young in two seasons.

The pied colouring in this mutation appears to be mainly restricted to the chest, belly, primary flights, tail, back of head and neck. Almost total absence of pied markings on the back, mantel and wing coverts (at this stage of development at least) suggests a "harlequin" pied mutation.

It seems reasonable to assume that this mutation will be established.

Australian Secondary Mutations.

PIED CINNAMON See Plate No. 11D

A Pied Cinnamon Elegant Parrakeet was bred by Hank Jonker in 1990, from a Western Australian Cinnamon male and a normal female which were both split to the recessive Pied mutation. A pied cinnamon and three cinnamons were reared by this pair.

Overseas Mutations

LUTINO

A Lutino mutation which is reported to be of an autosomal (non sexual chromosome) recessive inheritance appeared in Belgium in 1972 and is now well established.

The fact that this mutation has very pale blue wing bands and frontal band suggests to us that this is a very advanced cinnamon mutation which would explain its recessive inheritance. All well advanced cinnamon mutations, i.e., nearing lutinoism, appear to be of autosomal recessive inheritance.

In a true Lutino mutation all blue coloured areas of the normal bird is replaced with snow white and the inheritance is sex linked recessive.

CINNAMON

The European Cinnamon mutation of this species is similar in appearance to the Australian Cinnamon. It is said to have a recessive inheritance.

This mutation first appeared in 1982 but its origins are unclear.

The Pied mutation appeared in the aviaries of Herr Buckholz in Germany during 1978. It is a dominant mutation in which the pied areas usually increase with each successive moult. The beak and legs are of a light horn colour.

Hybrids

Hybrids have been recorded between this species and the Rock Parrakeet (*Australian Aviculture* 1952), Blue-winged Parrakeet, Turquoise Parrakeet and Scarlet-chested Parrakeet (Edward Boosey, *Keston Bird Farm, UK. Cage Birds* 23rd July 1953).

BLUE-WINGED PARRAKEET

Plate 12. Blue-winged Parrakeets (pair) male on left.

BLUE-WINGED PARRAKEET

(Neophema chrysostoma)

Derivation:
Neophema:
    Neo — from *Neos,*
      Greek for new
    Phema — from *pheme,*
      Greek for voice
chysostoma:
      chrysos — Greek for golden
      stoma — Greek for mouth

Description:

Length 21 cm
Weight 54 g average,
    females usually smaller
Appearance as in Plate No. 12

Classification

The Blue-winged Parrakeet, *Neophema chrysostomas,* is another less specialised member of the blue frontal banded, green bodied, group of Neophema parrakeets (formerly Neonanodes) from southern Australia including Tasmania.

The common name "Blue-winged" parrakeet refers to the broad, bright dark blue band which extends from the shoulder of the wing to the primary flight feathers and distinguishes this species from all other members of the genus. Other distinguishing features are, a less distinctive two-toned dark blue frontal band which does not extend beyond the eye; the yellow of the lores area extends beyond the eye and the body colour is a dull mid green.

Although this species has an extensive range its migratory and nomadic habits have prevented the formation of any physical or colour variations (i.e., subspecies).

Earliest Report

This species was first described by Kuhl in *Nova. Act. Phys. Acad. Leop. Carol.* Vol. 10, page 50, 1820 and illustrated on Plate 1 under the name *Psittacus chrysostomus.* The location was incorrectly given as Nova Hollandia instead of Tasmania.

Temminck described this species as *Psittacus venustus* in the *Transactions of the Linnean Society,* Vol. 13 page 121 London 1821, also with an incorrect location namely King Georges Sound.

After numerous genera and species names were bestowed on this species during nine decades Gregory Mathews gave it the present name of *Neophema chrysostoma* in 1910 (*Nov. Zool.,* Vol. 17 page 500).

Another early illustration was in Swainson's *Zoology Illustrated* Second series, Vol. 1, plate 21, 1829.

Range Habitat and Field Notes

The Blue-winged Parrakeet is a predominately migratory species with two distinct populations — Tasmanian and Mainland. They breed during spring and summer in the southern extremities of each population's range, then migrate northward during the autumn months to winter mainly in the northern inland regions of their range.

The Tasmanian population breed in Tasmania, then the majority fly across Bass Strait to winter in inland South Australia, Victoria and New South Wales. Whereas the mainland population breed in southern Victoria and south-eastern South Australia, then move north to winter in north-eastern South Australia, north-western New South Wales and south-western Queensland.

The extent of this northern movement is obviously dependent on prior rainfall and hence the availability of food. Permanent water supplies for stock, in the form of bores etc., means this species movements are far less restricted than in years gone by.

Blue-winged Parrakeets range over all of Tasmania, the larger Bass Strait Islands, King Island, all of Victoria, the extreme south-eastern tip of New South Wales as well as all of western New South Wales as far east as the Naron Lake District, then north-west to Cunnamulla and the Birdsville district of south-west Queensland and all of eastern South Australia from Eyre Peninsula to the Lake Eyre district.

This species may be found in a greater variety of habitat than any other Neophema, ranging from coastal salt marshes and dune areas, coastal heathland, tree studded grasslands, farming and grazing country, open forest, mountain heaths and forest, to in the north, the dry open scrubland and arid saltbush plains.

The extreme variation in habitat is reflected in a similar variation in diet. Blue-wings have been observed feeding on the seeds of numerous native and introduced grasses and herbaceous plants as well as vegetable matter, flowers, blossom, small fruits and some insects and their larvae.

Blue-winged Parrakeets have often been observed feeding in the company of Elegant Parrakeets and Orange-bellied Parrakeets.

Stan first observed this species in the field on the northern coast of Tasmania during January 1972, foraging on low lying sandy grassland which extended from the coastal dunes inland for approximately 1 km before the commencement of low scrubland. A flock in excess of 100 birds, consisting of adults and juveniles, moved silently around the low grasses, feeding mainly on dandelion seeds.

Periodically single birds or pairs would rise almost vertically from the feeding ground to in excess of 50 m high, then fly to the taller forest country beyond the scrubland, approximately 2 km away. These birds were obviously returning to their nest sites to feed their young.

As these birds rose from the ground they uttered a sharp whistling *ssit ssit* and continued to call until out of sight.

Our most recent sighting of Blue-wings was in June 1987 when in the company of Margaret Cameron, Andrew Isles and Norman Wettenhall we were driven to Werribee Sewerage Treatment Works and Farm to search for Orange-bellied Parrakeets. Unfortunately our only sighting of Orange-bellieds that day were of a few birds flying high above us and were unidentifiable as far as we were concerned, although our far more experienced colleagues assured us they were Orange-bellieds.

During our frantic search for the Orange-bellied we sighted a small group of Neophemas flitting around a swampy area which could probably be best described as salt marsh. Closer examination proved these birds to be Blue-wings. They fed mainly on the ground and would allow a reasonably close approach before rising to fly 100 m or so to land and feed again. We were surprised that they flew quite high to cover such open country. On occasions an odd member of the group would land on the top of one of the small shrubs that dotted the area to survey the surrounds. Blue-winged Parrakeets cover short distances with a floating, almost uncertain flight but long distances are undertaken with a fast, high and direct flight. When landing the tail feathers are fanned.

Breeding in the Wild

Breeding usually occurs from October to December and at times extends to February in Tasmania.

Nest sites may vary from hollow stumps or fence posts close to ground level to hollow limbs or holes in the trunks of trees, often at great heights. The females of this species appear to carry out the majority of the nest hollow preparations prior to breeding.

Four to six white rounded oval eggs are laid on a base of rotten wood dirt, usually at two-day but sometimes three-day intervals. The female usually incubates for 20 days before hatching occurs and the young fledge about 30 days later.

After breeding, Blue-wings usually form family groups or small flocks which are retained for many weeks often until the northern migration commences.

Aviculture

Probably the world's first breeding of the Blue-winged Parrakeet is that recorded by Hopkinson in his *Records of Birds Bred in Captivity,* 1926, of Cornely's success prior to 1885 (*teste Bull* 1885–86, page 563). Neunzig states there were a few other successes in Europe about the same time.

Dr H. D. Gröen in his *Australian Parrakeets* states this species was first bred in France in 1879 but fails to supply details.

H. Zomer in his *Australian Grassparrakeets and their Colour Mutations* 1989 states that they were first imported into Europe in 1874 and were bred in a zoo in Berlin, Germany as well as France in 1879, but also fails to supply details.

The first breeding in the United Kingdom was by F. R. Facey in 1909 for which he received an Avicultural Society Medal (*Avicultural Magazine* 1909, Page 357 and 1910 Page 198).

Australia's first recorded breeding of the Blue-wing was by S. Harvey of Adelaide South Australia in 1935 and he was awarded a Bronze Medal by the Avicultural Society of South Australia for his achievement. Adelaide Zoo bred this species in the 1936–37 season.

Our first breeding was recorded in Stan's aviaries in 1963 when one pair of a small colony (five birds), fledged three young. They were housed in an aviary 3.6 m (12 ft) long, 1.8 m (6 ft) wide and 2.1 m (7 ft) high, with half the length being a fully roofed shelter.

Stan's records also state that in 1964 only one young was reared by this colony. In 1965 in the now increased colony No. 1 pair reared four, No. 2 pair reared three and pairs No. 3 and 4 lost their young during a heat wave. During 1966 this colony reared 10 young. Double brooding has been recorded often in this colony.

Due to the southern breeding range of the Blue-wing (no further north than 34°S) nestlings of this species do not withstand heatwave conditions at all well. This is especially the case after the pin feather stage is reached.

To help combat this susceptibility to heat, housing in larger well insulated and shaded aviaries as well as the use of large, well ventilated nest boxes with lids capable of being propped partially open on hot days is advisable. A day temperature over 35°C (95°F) during the critical stage from pin feather onwards will surely result in disaster if precautions are not taken.

Stan has always colony bred Blue-wings in large rather open aviaries with good results and believes this species is the only totally successful colony breeder of the genus. Yet we both believe that pairs housed individually in sheltered aviaries three to four metres long, about 1 m wide and 2 m high will give more consistent breeding results. The mortality rate is also reduced in juveniles when bred and housed in these smaller aviaries as opposed to the colony method.

Because of this Neophema's colony breeding potential we have never found it necessary to house Blue-wings with other parrot species but have successfully housed and bred them with a number of species of native pigeons.

The following information on the flock breeding of Blue-wings was provided by Allan Hogan from Sydney's south-west.

Allan's two breeding pairs were housed in aviaries 2.4 m (8 ft) long, 1.6 m (5 ft 4 in.) wide and 2 m (6 ft 6 in.) high, each with a pair of Bourke's Parrakeets and a pair of Turquoise Parrakeets.

The birds were fed on a diet of canary mix and finch mix with a little sunflower seed and hulled oats. Sprouted sunflower seed and red millet were provided every afternoon and either endives, spinach, chickweed or chopped up apple were fed daily. Rain water was provided for drinking with Calcium Sandoz and Avi Vit added twice a week.

During the 1988–89 season a 12-month-old pair laid three eggs early in November, hatched them late in November and fledged three young late in December. Another clutch of four eggs were laid early in January 1989, four young hatched, but only two fledged towards the end of February. Two died in the nest.

In the 1989–90 season this pair fledged three young in mid November. Then four eggs laid in another box in late November were fostered under Turquoisines who hatched, then fledged four young in early January. A third clutch of eggs was laid early in December and three young fledged late in January 1990.

A second pair (12 months old) fledged three young late in November 1989 and four more in February 1990, providing a magnificent total of 17 young from two pair in one season.

Blue-wings are a docile inoffensive species which can be housed with finches, small soft-bills and doves etc. Care should be taken not to house them with other species of Neophemas or larger parrot species which may be aggressive towards them. Allan Hogan's example is, in our opinion, the exception rather than the rule with flock breeding.

Jim has successfully housed and bred both Gouldian Finches and Blue-wings in the one large aviary 5 m (16 ft) long 1.8 m (6 ft) wide and 2.4 m (8 ft) high. There were several pairs of Gouldians and one pair of Blue-wings and they fledged young at the same time during 1988.

We have found that Blue-wings will do more damage to plant life in a planted aviary than any other species of Neophema.

Our own results indicate the average number of young reared in a clutch is four although we have on occasions had five fledge from the one nest. Mick Grixti from the south-west of Sydney had a nest of six fledge in 1989. This pair were housed by themselves in a conventional small Neophema aviary and add credance to our belief that they will do even better when housed this way.

Sexing

The adult male is generally of a brighter colour, particularly the blue frontal band, the yellow area between the beak and eye (lores) and the broad blue wing band. All males have a much brighter, more extensive and deeper coloured yellow belly and most have a small patch of yellow orange in the centre of the abdomen between the legs. This spot is very occasionally carried in a dull, reduced form by the female. The green of the chest area extends further down in the adult female.

Immatures are similar to adult females, only duller with large black eyes, little or no blue frontal band and a less extensive blue band on the wings. An underwing stripe is present in most immatures, Sexing of immatures is extremely difficult. Adults during their moult are also extremely difficult to sex.

Display

The male advances towards the female in an excited manner, stretched to his full height and uttering a soft twittering whistle. This display often results in the male feeding the female.

130

Nests

Blue-wings themselves are not difficult to satisfy with nesting sites. Hollow logs or nest boxes, either vertical, partially inclined or horizontal may all be accepted. However, only larger logs or boxes should be supplied if the young are to survive the excessive heat of summer on most of the Australian mainland. Thankfully, and unlike most Australian parrots, Blue-wings often show a preference for a larger nest site.

If a hollow log is to be used for this species we would suggest it have a minimum internal diameter of 25 cm (10 in.) and a length of about 50 cm (20 in.) hung either vertically or partially inclined.

Our own preference is for a nest box approximately 30 cm (12 in.) cube, with a 6.2 cm (2.5 in.) entrance hole near the top (see Figure 6C) with 6 mm (0.25 in.) ventilating holes drilled on opposing sides to provide cross ventilation. Nest sites should always be situated under shelter, out of direct sunlight in the coolest possible position and filled to a depth of 5 cm (2 in.) with any of the recommended nest fillings.

Nesting and Hatching

This species' intention to breed is marked by displays and increased interest in the chosen nest site, particularly by the female who spends long periods in the nest.

In our rather severe climate of the south-western suburbs of Sydney we have recorded eggs laid between October 10 and December 13.

The average clutch of four or five white eggs are laid at usually two or sometimes three-day intervals. Occassionally clutches of six or seven eggs have been recorded.

Incubation is carried out by the female and usually commences with the laying of the second or third egg. Incubation periods of 19 and 20 days have been recorded.

The chicks hatch with a silvery white down, the eyes open at eight or nine days, pin feathers are visible at 10 days, flight and tail feather development is well advanced at 18 days and the chicks are fully feathered at 28 days of age. We have recorded fledging periods of 31 days to 35 days.

Fledglings leave the nest with the appearance of a dull adult female, with little or no blue frontal band, large black eyes which lack iris rings and a yellow beak.

Immatures moult into adult plummage at about six months of age and can breed at 12 months.

A Brooding hen with newly hatched chicks.

B Seven-day-old nestlings with two late hatchlings.

C Twenty-day-old nestlings.

D Thirty-day-old nestlings just prior to fledging.

E Blue — a new mutation.

Plate 13. Blue-winged Parrakeets — nestlings and mutation.

Mutations

BLUE See Plate No. 13E

A true blue mutation, devoid of all yellow colour pigment has been bred by Max Peek of Murray Bridge South Australia, during the 1991 breeding season. The blue youngster and a normal nest mate were reared from a normal coloured pair who were brother and sister.

The green areas of the bird have been replaced by a blue described to us as the blue of a Blue Princess Parrot. The normal blue areas are retained in the same depth of colour. All yellow areas have been replaced with white.

We would expect this mutation to be of recessive inheritance as all blue mutations appear to be.

Blue mutants of this species were reported in England about 1970 but apparently the strain was lost. H. Zomer *Grass Parrakeets and their Colour Mutations* 1989 p. 93.

We can find only one other record of a mutation occurring in this species, either in Australia or overseas and that is of a pure yellow specimen observed by an ornithologist in a small flock in the wild. Of course we can only speculate as to whether this bird was a Lutino, Dilute Yellow or Cinnamon Yellow. (*Foreign Bird Keeping,* Edward J. Boosey 1950 page 141).

A colour variant strain in which the orange belly patch is extended considerably has been produced by selective line breeding in a similar manner to Red-fronted Turquoisines and Red-bellied Scarlet-chesteds.

Hybrids

Hybrids have been recorded between this species and the Elegant Parrakeet (England — Edward Boosey *Cage Birds* 23/7/53. Australia — *Australian Aviculture* 1952 page 86) and with the Rock Parrakeet (*Australian Aviculture* 1949 page 106).

It is interesting to note that Edward Boosey of Keston Bird Farm England listed the Blue-wing × Elegant Hybrid as having reproduced at Keston. If this was true it is the only record we can find in the literature of a fertile Neophema hybrid. We have many verbal reports of fertile hybrids between the blue browed species (Neonades).

133

ORANGE-BELLIED PARRAKEET

Plate 14. Orange-bellied Parrakeet (male).

ORANGE-BELLIED PARRAKEET

(Neophema chrysogaster)

Derivation:
Neophema:
    Neo — from *Neos,*
      Greek for new
    Phema — from *pheme,*
      Greek for voice
chrysogaster:
    cryso — from *chrysos,*
      Greek for golden
    gaster — Greek for belly

Description:

    Length 21.5 cm average
    Weight 50 g average
    Appearance as in Plates 14 and 15

Classification

The Orange-bellied Parrakeet is currently the rarest and the second most specialised member of the Neonanodes group as well as of the Neophema genus as a whole.

Even in the past when all Neophemas enjoyed a greater and more populous range, it is probable the Orange-bellied Parrakeet was not as numerous as other members of the genus. Their range may have extended from the vicinity of Adelaide, South Australia, through all suitable habitat along the southern and eastern coastal areas to the Sydney region of New South Wales as well as much of Tasmania.

This species is easily distinguished from the other members of the Neonanodes group by the bright grass-green colouring of the upper body as well as by the distinctive buzzing *zzit zzit* alarm call which is unlike that of any other Neophema.

Like the Blue-winged Parrakeet, this species is migratory, wintering on mainland Australia then returning to Tasmania during October and November to breed, then commencing their return to the mainland during March. Apparently not all birds migrate, for there have been isolated sightings on the mainland during summer and in Tasmania during winter.

Plate 15. Orange-bellied Parrakeet — Female with youngsters at the entrance of nesting hollow.

We would suggest that the apparently isolated and sedentry population which was recorded breeding in the Sydney region prior to, and just after the turn of this century, was a remnant of a once far more extensive range of this species. Alfred J. North in his *Nests and Eggs of Birds Found Breeding in Australia and Tasmania* Volume 3, Page 160, 1912, recorded sightings at Middle Head, Long Bay, Penshurst, Blacktown and Riverstone, which are all suburbs of Sydney, during the late 19th century and early 20th century.

Gregory Mathews separated the South Australian form in his *A list of the Birds of Australia* page 138, 1913, as *Neonanodes chrysogaster mab*, which he felt had a broader and more pronounced blue frontal band. Then in his *The Birds of Australia* Volume 6, page 441, 1916–17, he expresses his concern for this subspecies and its pending extinction. The authenticity of this subspecies in a migratory species was always dubious and is currently not recognised.

Earliest Report

This species was first described by Latham in his first *Supplement to the General Synopsis of Birds* 1787, page 62, as the Orange-bellied Parrakeet. The specimen information was supplied by Mr. Pennant who had apparently obtained several specimens which were taken in Tasmania on Cook's third voyage.

Later Latham proposed the scientific name *Psittacus chrysogaster* in his *Index of Ornithology* 1790, page 97.

The first illustration appears to be that in Gould's *Birds of Australia* Volume 5, Plate 39, 1848, listed as *Euphema aurantia*. It was also illustrated in Diggles *Ornithology of Australia* Volume 2, Plate 81, 1877 as *Euphema aurantia*.

In 1891 the name in current use, *Neophema chrysogaster* was applied by Salvadori in *Catalogue of Birds in the British Museum* Volume 20 page 573.

Range, Habitat and Field Notes

The Orange-bellied Parrakeet is a migratory species which arrives in Tasmania during spring and leaves again for mainland Australia in the autumn.

In Tasmania this species is currently confined to the south-western half, although historically it covered a far greater range. Gould observed Orange-bellied Parrakeets in small numbers in the neighbourhood of

Hobart and New Norfolk and in greater abundance on the Actaeon Islands at the entrance of the D'Entrecasteaux Channel, during his visit of 1839. Littler in his *Birds of Tasmania* 1910, page 98, records a sighting of eight birds on Ninth Island off the north-east coast of Tasmania. These birds were probably on migration to Tasmania, during late September 1909.

Today, sightings mainly occur in coastal areas of western and southern Tasmania with isolated inland records, often at high altitudes. During migration this species passes through the Hunter Islands Group and King Island, where some birds spend the winter months.

On mainland Australia, Orange-bellied Parrakeets currently range through the coastal regions from the Port Albert and Gippsland districts of Victoria in the east to the Coorong and occasionally the Lake Alexandria areas of South Australia at the western limits of their range. There is also an isolated sighting at the I.C.I. Saltworks, north of Adelaide, South Australia in these western limits.

The Orange-bellied Parrakeet, when on the mainland, is found in saltmarsh areas (particularly around Port Phillip Bay in Victoria), coastal heathland and low lying grasslands as well as in coastal sand dunes (particularly in South Australia).

In Tasmania they mainly inhabit grasslands, open scrubland in coastal areas, sedgeland dotted with clumps of buttongrass and buttongrass plains.

The diet consists of the seeds of native and introduced grasses and weeds as well as the seeds, fruits, flowers and leaf tips of shrubs and the flora of tidal flats, saltmarsh, coastal heathlands, sand dunes, low lying grasslands and sedgelands.

This parrakeet spends most of its time feeding on or near the ground and when not feeding will often shelter under low lying foliage. Feeding has been observed throughout the day as well as before dawn and after dusk.

They are usually observed in small groups, pairs or singly and close approach is difficult. When flushed they rise quickly while uttering their distinctive buzzing alarm call, then after reaching a good height they descend and fly off just above the vegetation level.

Orange-bellieds have been observed drinking daily from regular watering places at Point Wilson, Victoria, where groups of up to 10 birds arrive and perch near the water supply for several minutes before descending to drink, one by one. At times these birds would drink from pools of rainwater. (*Ecology of Orange-bellied Parrots*, Lyon, Lane, Chandler and Carr, *the Emu*, Volume 86, Part 4, Page 198).

The same study revealed regular roosting areas in scrublands that range from 1 to 3 m high.

Unlike our first attempt to observe Orange-bellied Parrakeets in the field, as mentioned elsewhere, our second attempt was an almost instant success. On the morning of September 4, 1987 Andrew Isles had driven us to Port Wilson, Victoria and parked his car at the commencement of "The Spit". We started to walk along the elevated access road which runs parrallel to the beach at this point.

Barely 50 m had been covered, when an adult pair was sighted, feeding on the top of sea heath *(Tranfenia pauciflora)*. The very obvious pair sat quietly on top of the heath, nibbling the foliage tips or seeds and giving us perfect viewing with the aid of our binoculars and Andrew's telescope. Another 100 m along the roadway and a single bird was spotted, with the aid of binoculars. This bird also fed quietly on top of the heath. We all agreed its slightly duller colouration and smaller head was indicative of an adult female.

Just a little further along the dirt road two more Orange-bellieds were observed, feeding, only this time they were both juveniles, identifiable by their generally duller appearance.

The bright grass-green body colour of this species, even in juveniles, makes it readily identifiable in the field from the other, dull green coloured members of the Neonanodes group.

Their flight reminded us more of a budgerigar in flight than of flight of another species of Neophema.

That morning we enjoyed bird watching at its best, with good, clear, identifiable sightings, all without moving off the roadway or getting our feet wet. In under an hour we were back in Andrew's car heading towards Melbourne with a flying start to our busy schedule.

The particularly sad and depressing feature of this species is its decline in numbers from numerous sightings of thousands late last century to a current total population of probably less than 200 birds.

Destruction of suitable wintering habitat along the coast of south-eastern Australia and to a lesser degree, habitat destruction in Tasmania is the direct and major factor for the demise of this species. Illegal trapping for aviculture has also been cited as contributing to the decline.

In defence of aviculture we must point out that Orange-bellieds have always been extremely rare in Australian aviaries. However, it is fact that during the 1960's certain unscrupulous people trapped this parrakeet in the Coorong region of South Australia. They were then smuggled to Europe via Sydney, New South Wales. Trapping pressure has never been great and that pressure having been removed the species will recover if other environmental factors are satisfactory. In this species the environment is obviously not satisfactory.

On a brighter note, the establishment of a "Recovery Plan for the Orange-bellied Parrot" was a positive step. It was developed after three year's preliminary work on the species in the field by the Tasmanian National Parks and Wildlife Service in co-operation with the Fisheries and Wildlife Division of Victoria and the South Australian National Parks and Wildlife Service and many volunteers.

This Plan in turn led to the establishment of a "Recovery Team" which consists of representatives of the Tasmanian, South Australian and Australian National Parks and Wildlife Services, the Victorian Fisheries and Wildlife Service, and the Royal Australian Ornithologist's Union. The Recovery Team's first meeting was in 1984.

Initiatives already in place, due to the efforts of this Recovery Team, include:

1. The restriction of grazing of critical saltmarsh areas in Victoria;

2. The improved management and vermin control of a garbage tip site which is adjacent to Orange-bellied wintering grounds by the Kingston Council in South Australia;

3. The protection of vital dry saltmarsh areas;

4. The propagation of "fat hen" which has become an important food item of the Orange-bellieds in the Point Wilson sewerage farm evaporation paddocks by the Melbourne Sewerage Authority;

5. The refusal of a subdivision application to commercially develop an area of wintering habitat in south-eastern South Australia;

6. The abolition of grazing rights on the parrakeets feeding grounds along the west coast of Tasmania which are used during the migration of this species;

7. The establishment of a captive breeding program in Tasmania;

8. Sponsorship enabling constant monitoring of the species movements.

Of course only time will tell what the future holds for Orange-bellied Parrakeets but we must all be heartened by the positive steps taken by all the governmental departments.

Flight

When flushed this species rises quickly to a considerable height then flies off with a fast, floating motion which is interrupted by periods of gliding flight.

Breeding in the Wild

The Orange-bellied Parrakeet's breeding cycle is just as specialised as the rest of its natural history.

After wintering on the south-eastern coast of the Australian mainland this species commences its migration back to Tasmania and its breeding grounds, during September and October. Their route across Bass Strait passes over King Island then to north-western Tasmania.

The parrakeets move down the west coast, feeding as they go, to their breeding grounds in the extensive sedgelands of coastal south-western Tasmania from Birch's Inlet at the southern end of Macquarie Harbour, south to South West Cape and then east to the vicinity of Red Point.

Nesting occurs during October, November and December. Three to six white, rounded oval eggs have been recorded and are laid on a base of rotten wood in holes in the trunk or in hollow limbs of living eucalypt trees up to about 25 m (80 ft) above the ground. Early observers recorded nestings in hollow stumps and fallen dead trees.

Successful breeding pairs have been known to return to the same nesting hollow year after year.

Peter Brown of the Tasmanian Wildlife Department found the average depth of nest sites examined in Tasmania's south-west, was 45 cm (1 ft 6 in.) from the entrance hole to the nest base.

After breeding, during March and April, when the young are independent, the parrakeets move up the west coast to north-western Tasmania in preparation for their return journey to their wintering grounds on mainland Australia. The return route across Bass Strait which is via the Hunter Islands Group and King Island is carried out at a more leisurely pace than the trip south, with many birds lingering for several weeks on King Island where, occasionally, a few remain for the entire winter.

Aviculture

The Orange-bellied Parrakeet was almost unknown to English and European aviculturists of the last century. Apparently a pair was exhibited in London Zoological Gardens prior to 1903. (Seth-Smith *Parrakeets*

143

1903). Edward Boosey held one pair at the Keston Bird Farm, England, after World War II but was unsuccessful with them.

In Australia, Mr. Percy Peir of Marrickville, Sydney, New South Wales, in a letter dated 5th October 1909, writes of Orange-bellied Grass Parrakeets being numerous in the Sydney suburbs of Penshurst and Blacktown during the 1880's, when they were often found in the possession of bird-keepers. He also refers to a beautiful specimen exhibited at a Sydney Show during 1907, which was reported to have been captured at Riverstone (an outlying Sydney suburb), and of a hen he later received from the same area. (A. J. North *Nests and Eggs of Birds Found Breeding in Australia and Tasmania* Volume 3, page 160, 1912).

Alan Lendon in *South Australia Ornithology* 1948, page 24, states "Hamilton claims to have bred a solitary bird during the 1939–45 war, but did not report it." (Prestwich, *Records of Parrots Bred in Captivity*, Additions, 1954, page 66). This stock was probably the result of the report by Floyd in 1938; "In 1934 Dr. William Hamilton, an Adelaide aviculturist, captured ten birds near Robe".

During the late 1960's and early 1970's many Orange-bellied Parrakeets were exported illegally to Europe. As a result the first European breeding of this species was achieved by van Brummelen in Holland during 1971 when three youngsters were reared from a five egg clutch. Then in 1972 J. Postima, also of Holland, reared a single bird. (Rosemary Low, *Endangered Parrots*, 1984, page 102). There is also an unconfirmed successful breeding from West Germany at about the same period.

Unfortunately this species does not appear to have established in European aviaries.

The first authenticated and documented breeding of the Orange-bellied Parrakeet in Australia was the outstanding achievement by the late Fred Lewitzka of Adelaide, South Australia, in 1973. The following summary of this event was gleaned from accounts published in *Bird Keeping in Australia*, Number 23, pages 60–64, April 1980, and in *Australian Parrots*, Hutchins and Lovell, 1985.

Three pairs of adult birds were housed in individual aviaries 9.5 m (31 ft) long, 2.5 m (8 ft 2 in.) wide and 2.5 m (8 ft 2 in.) high, which included a 2.7 m (8 ft 10 in.) shelter. The flight areas were planted with various shrubs and seeding grasses.

During the five seasons preceding the successful breeding, all pairs had produced varying clutches of infertile eggs which were always laid in December and January. Increased activity and moonlight night calling was observed from October until breeding commenced.

The successful pair was housed with four Red-faced Parrot Finches, five Hooded Siskins, two Scarlet-chested Parrakeets and two Red-backed Button Quail. The male Orange-bellied Parrakeet had been an occupant of this aviary for six years and the female for three years.

The chosen nest site was a natural hollow log 43 cm (17½ in.) long, an internal diameter of 10 cm (4 in.), with both ends closed and a 5 cm (2 in.) diameter entrance hole near the top. The log was hung in the open flight at a 45 degree angle, 1.8 m (6 ft) above ground level and contained a mixture of sawdust, dirt and fine peatmoss for nesting material. Protection from adverse weather conditions was provided above the log.

Both male and female were seen to chew around the entrance hole of the log before the first egg was laid on January 14, 1973, followed by three more laid at two-day intervals. The incubation was carried out by the female who usually left the nest during the late afternoon when she was fed by the male, before she visited the food tray and water dish. On other occasions the cock would feed the hen at the entrance of the log.

After 12 days incubation the nest was inspected and three eggs found to be fertile. The first chick hatched on February 9, followed by the second on the 11th and the third chick on February 12, but which died the same day. Chicks hatched with an off-white down and the incubation period was determined at 21 days. Most of the feeding of the young was carried out by the female, who showed a distinct preference for seeding grass heads. Other foods available were panicum, white millet, Japanese millet, canary seed, crushed sunflower seed and niger seed.

Fledging of the first chick occurred on March 17 followed by the second chick the next day after spending 36 days and 35 days respectively in the nest. The youngsters were fed by the female only for the first few days, then by both parents. They were seen pecking at seeding grass and dry seed on the ground only one week after fledging and considered independent four weeks after leaving the nest.

In March 1974 another youngster was fledged by the same pair and previously unpublished information found in Fred's records by his son John, indicates that in February 1975 two more young Orange-bellieds were fledged by this pair.

About this period the South Australian Wildlife Authorities directed Fred Lewitzka to release all of the Orange-bellied Parrakeets into the wild. One pair was allowed to go to the Adelaide Zoological Gardens and the rest were released.

This type of directive was typical of Australian Wildlife Departments' mentality during that period. Hopefully such an achievement would be met with a more positive response today.

In 1979 a research program for the Orange-bellied Parrakeet was supported by the World Wildlife Fund. A flow-on from this initial research was the decision of the Tasmanian National Parks and Wildlife Service to construct an aviary complex in a new wildlife research centre at Green Point Nature Reserve near Hobart.

The purpose of the construction of this aviary complex was to conduct a trial breeding program with two closely allied species, the Rock Parrakeet, *Neophema petrophila* and the Blue-winged Parrakeet, *Neophema chrysostoma*.

The aviary complex, which was completed in 1982, consisted of two blocks of five aviaries each backing on to a service area. Both groups of aviaries have removable partitions which allows an aviary size to increase up to five times. Each aviary is 7 m (22 ft 9 in.) long and 4 m (13 ft) wide with a curved roof which provides a varying height from 1.5 m (4 ft 11 in.) at the front, to 3 m (9 ft 9 in.) at the rear. One-third of each aviary is roofed with insulated colorbond corrugated iron sheeting as well as having an inner sheltered feeding area of 1.5 m (4 ft 11 in.) × 1.5 m (4 ft 11 in.). All aviary floors are of blue metal gravel and both natural and sawn timber perches are provided. The aviary complex has a rear safety corridor to prevent escape.

This captive breeding program was placed under the control of Peter Brown of the Department of Lands, Parks and Wildlife in Tasmania, who had been involved in the Orange-bellied research project from the beginning. Peter has an aviculture and zoo background in the United Kingdom and was a zoo curator for 12 of his 18 years experience in that area. He was a council member of the English Avicultural Society, the World Pheasant Association and the Federation of Zoological Gardens of Great Britain and Ireland.

The breeding program of the Rock Parrakeet and Blue-winged Parrakeet, which were both flock bred, commenced during the 1982–83 season and proved to be highly successful, eventually leading to the release back into the wild of 60 Rock Parrakeets and at least 24 Blue-winged Parrakeets.

The stage was now set for the implementation of the captive breeding program of the Orange-bellied Parrakeet. Support for the program was given by the Orange-bellied Parrot Recovery Team at their meeting in South Australia in June 1985.

Ten juveniles were captured late the following February (1986) in south-west Tasmania prior to their migration. They were housed in two flight cages 2 m (6 ft 6 in.) × 1 m (3 ft 3 in.) × 1 m (3 ft 3 in.) which were placed in a room within the aviary complex. Soaked and dry seed was offered and a broad based antibiotic (terramyacin) given in the drinking water.

After one month all the birds were released into one of the single compartment aviaries where they settled in well. Then, as the winter approached many started to moult in yellow feathers, which were first thought to be the result of a vitamin deficiency. As the condition worsened and medications had no effect, it became apparent the problem was far more serious. The end result was the diagnosis of Psittacine Beak and Feather Disease and the death of seven Orange-bellieds.

The remaining three birds, which were two females and one male, appeared healthy so were placed in a double aviary where the cock was seen displaying to the females during September. Natural nesting logs 60 cm (2 ft) to 80 cm (2 ft 8 in.) long with natural entrance spouts or holes near the top were hung in the shelter about 2.5 m (8 ft 2 in.) high.

On October 28, 1986 one hen was seen at the entrance of a vertical log and two days later she refused to leave the log. By November 15, the other hen was seen at the entrance of a log and two days later was remaining in the log for long periods. No attempts were made to inspect the nest sites during the breeding period.

From the end of November the first female came off the nest regularly and on December 5, at least two chicks could be heard calling in the log, then on December 30, a perfectly feathered youngster left the nest, another fledged on January 3 followed by a third on January 6.

On January 7, young could be heard in the second nesting log. A youngster fledged on January 26 and upon inspection two dead chicks were found in the nest. The male was never seen to assist this female with the rearing. All four young were reared to independence.

Six new birds were captured on March 5, 1987 and were acclimatised and housed in a separated group of aviaries to isolate them from the possible infection of Psittacine Beak and Feather Disease.

During early April the aviary bred young started to moult in yellow feathers which eventually resulted in the death of two birds with Beak and Feather Disease. A third bird was killed in an accident, while the fourth bird returned to normal colouring and apparent good health.

Eight young were reared during the 1987–88 season, seven from two pairs of the newly captured, 12-month-old birds and one from the original stock.

An attempt to foster six Orange-bellied eggs in two Rock Parrakeet nests during the 1988–89 season ended in failure, although 22 young Orange-bellieds were reared naturally that season.

During 1989 it was decided to relocate the breeding aviaries to a new locality — a task which took longer than anticipated and resulted in the birds being placed in their breeding aviaries too late in the season. Nevertheless three young were reared in the 89–90 season.

The 1990–91 season was more productive with 14 youngsters fledging, all of which survived their juvenile moult and were still alive and well in June 1991.

A pilot release into the wild commenced in the early summer of 1991 with a group of mixed aged, aviary bred birds, which were identified by coloured leg bands. The release aviary was erected on the breeding grounds in south-western Tasmania and the birds placed in it by late September. Their diet was changed to a totally natural one before release. The release was effected by removing a panel from the aviary wall allowing the birds to leave and return as they please.

The breeding and release area is closely monitored hence the freed birds are under close surveillance from the onset to assess their assimilation into the wild population and natural habitat.

This pilot release is aimed at establishing the ideal age for liberation, whether the birds will migrate with the wild population or not; if they migrate will they survive the trek and the winter on mainland Australia. During this project, incubation periods of 22 to 24 days and fledging periods of four to five weeks were recorded.

The diet used for the Orange-bellieds during this project consists of a normal canary mix to which is added extra millet, linseed, niger and rape seed. Natural greenfood is fed regularly and saltmarsh flora is fed once a week to provide the necessary salt required in the diet of this species. During the spring and the approach of the breeding season, the vitamin and mineral supplement SA 37 is added to the seed mix.

Peter Brown's final comment to Stan of this significant project under his control was that the Orange-bellied Parrakeet is not a subject for aviculture and it is unlikely it ever will be.

Since writing this chapter the first Orange-bellied release has been put into motion. Reports from the release area suggest an incredible degree of success with some of the birds. Almost certainly three hens and probably one male from the released stock actually bred with wild Orange-bellieds within a couple of months of the release.

After the breeding season the birds appear to have commenced the annual migration with the wild population.

To add further to the overall success of the program, 18 captive bred Orange-bellied were fledged on the first round and 11 on the second round during the 1991–92 breeding season.

Our experience in keeping Orange-bellied Parrakeets is limited to the keeping of one pair and then a single bird in Stan's aviaries during the 1960's.

It all started when Stan and his old friend and master aviculturist of the time, the late Bill Bond, were given the opportunity to select two pair of Orange-bellied Parrakeets from a group of 38 in the possession of a leading Sydney bird dealer of that period.

Stan and Bill had heard rumours of birds being trapped in South Australia and of the huge predominance of males amongst these birds, so they expected females to be hard to find.

The 38 birds, which were all in good condition, were systematically culled and only four possible hens were found. Of these four birds, two were probable hens while the other two were perhaps, young males.

Two nice males were selected to go with the probable hens, then Bill and Stan purchased a pair each. Unfortunately we believe the remainder of this group were smuggled to Europe.

Both pairs were housed in small Neophema type breeding aviaries and fed, our then basic seed diet of canary seed with a little millet and sunflower seed. Soon both pairs became so lethargic and obese that radical measures had to be taken.

Bill and Stan both had much of their aviary complexes opening into large planted aviaries, which in Stan's case housed softbills, native pigeons and a few Neophemas. Stan removed the Neophemas and released his Orange-bellieds into this large aviary which was 9 m (40 ft) long,

3 m (10 ft) wide and 2.4 m (8 ft) high, with reservations and concern for their well-being, in such a large aviary. Bill released his pair into similar conditions and within a few weeks his male was found dead with a head injury.

Stan's pair lost weight, became far more active and appeared to flourish, but a few months later his male met the same fate as Bill's.

Orange-bellieds were never sold on the open market in Sydney again so the lost birds could not be replaced. Stan's female survived for another year or so before it was also lost with head injuries. Dissection proved this bird to be a female.

The remaining hen was eventually brought into Stan's collection after Bill's sad and untimely death from lung cancer. This bird lived on for many years in a medium sized holding aviary fed on Stan's new basic seed diet of mixed millets and greenfoods. When it eventually died of natural causes the body was given to and mounted by our friend Charles Attard who found it to be in fact a poorly coloured male.

Stan has always regretted that this species came into his hands before his experience and expertise were equal to the task. Given the opportunity he would now hold this species in a 4 m (13 ft) to 5 m (16 ft) long, sheltered aviary and restrict the maintainence diet to dry and sprouted millets plus plenty of greenfoods.

Sexing

The adult female is noticeably duller than the male, has a much duller and reduced frontal band which lacks the paler blue upper edging of that of the male as well as a duller and often smaller, orange belly-patch.

Immature birds of this species have more pronounced sexual differences than the other members of the Neonanodes group. We had no difficulties in distinguishing the sexes of immature birds observed at Point Wilson, which would have been approximately eight months old. Fred Lewitzka also noted the pronounced sexual differences in fledglings. Young males are brighter than young females.

Display

Peter Brown provided us with this interesting description of this species display which he first observed in the field and later confirmed in captivity. The male stretches high, squares the shoulders of the wing, then extends the feathers of the orange belly-patch at right-angles to the body which serves to accentuate the bright orange patch and attract the females attention.

Captive Breeding Techniques

The successful breedings of this species, up till now, have been achieved by the use of large aviaries and in some instances, colony breeding which help combat the lethargic behaviour and tendancy towards obesity.

We would suggest that perhaps single pairs housed in adjoining, medium sized aviaries (4 m to 5 m long), with wire partitions, might eliminate the stress and injuries often common in very large aviaries, yet still provide the stimulation of a colony situation.

The use of a reduced non-fattening seed diet with plenty of greenfood should also combat obesity in this species just as it does with Rock Parrakeets.

Nests

Orange-bellied Parrakeets have been recorded to nest in both vertical and horizontal hollow logs about 45 cm (1 ft 6 in.) long with a 10 cm (4 in.) to 15 cm (6 in.) internal diameter, with both ends closed and an entrance hole of about 5 cm (2 in.) near the top of the log. Nesting logs hung high in the aviaries and under shelter are usually preferred, but if fixed in the open flight it must be protected from the weather.

Nest fillings of rotten wood or a mixture of sawdust, dirt and fine peatmoss have been used successfully. Particular care should be taken on mainland Australia to keep the nest site cool.

Nesting and Hatching

Within a few days of the selection of a suitable nest site the female Orange-bellied will disappear into the nest and apparently lay her first egg almost immediately.

Egg laying in captivity has been recorded from October to January. Three to five rounded-oval white eggs are laid, usually at two-day intervals, and are incubated by the female. The only accurately recorded incubation period available is 21 days.

The chicks hatch with an off-white down and fledge about five weeks later as dull versions of the adult female with only a trace of the blue frontal band, a reduced orange belly-patch, an orange-brown beak, and large black eyes. About three months after fledging the beak changes to dark brown and by nine months old they have moulted into sub-adult plumage.

Mutations

We can find no record of a mutation occurring in this species.

Hybrids

A hybrid has been recorded between this species and the Rock Parrakeet. (*Australian Aviculture* 1952, page 86.)

ROCK PARRAKEET

Plate 16. Rock Parrakeets (pair) male on right.

ROCK PARRAKEET

(Neophema petrophila)

Derivation:
Neophema:
  Neo — from *Neos,*
  Greek for new
  Phema— from *Pheme,*
  Greek for voice
petrophila: petro from *petros,*
  Greek for rock
  phila from *philos,*
  Greek for fond of

Description:

  Length 22 cm average; our
  observations indicate
  males appear marginally
  larger
  Weight 52 g average
  Appearance as in Plate 16

Classification

The Rock Parrakeet, *Neophema petrophilia* is the most specialised member of the green bodied blue frontal banded group (Neonanodes) as well as of the Neophema genus as a whole.

This specialisation involves its restriction to harsh coastal habitat which in turn imposes a basically terrestrial lifestyle and the need to nest on the ground.

The Rock Parrakeet derives its name from the habit of frequenting and nesting on rocky shore lines and offshore islands.

Distinguishing features of this species are the broader, bulkier body size (the largest member of the genus) and generally duller colouration. It is the only member of the Neonanodes group without yellow lores (the area between the eyes and the beak) the yellow being replaced by a blue wash on the lores and cheek area.

An eastern and western population is separated by an extensive section of the Great Australian Bight which lacks suitable habitat. This separation into two long-term, isolated populations has not resulted in the formation of subspecies or even noticeable physical or colour variations.

Gregory Mathews described a subspecies from Sir Joseph Banks Island in the Spencer Gulf, South Australia, which had a darker blue frontal band and a generally duller colouration. Later this subspecies was wisely ruled invalid.

Earliest Report

This species was first described by Gould in the *Proceedings of the Zoological Society* (London) 1840 Part 8, page 148, 1841, as *Euphema petrophila,* from specimens collected in Western Australia.

In 1891 Salvadori listed this species as *Neophema petrophila* in the *Catalogue of Birds in the British Museum,* Volume 20, page 574.

The first illustration appears to be that in Gould's *Birds of Australia,* Volume 5, Plate 40, 1844.

Range, Habitat and Field Notes

The Rock Parrakeet is a relatively stationary species in which there have been no population migrations recorded, unlike the closely allied Blue-winged and Orange-bellied Parrakeets.

Currently it is accepted that the range of this species is divided into an eastern and western population by approximately an 800 km stretch of coastline in the Great Australian Bight from the Denial Bay region of South Australia to the locality of Point Culver in Western Australia. This vast expanse of coastline is deemed unsuitable habitat due to the absence of cliffs, rocky areas and rocky offshore islands.

In contradiction there is one recorded sighting from this region by W. B. Alexander at Hampden Range about 190 km west of Eucla where the once coastal cliffs are now several kilometres inland, (*Birds of Australia*, G. M. Mathews, Volume 6, page 455, 1916). The lack of further sightings from this remote region has led to this observation being discounted.

The eastern population of the Rock Parrakeet extends from the Cape Jaffa and Coorong region of South Australia westward along the coastline including Kangaroo Island and suitable offshore islands to the vicinity of Denial Bay, South Australia.

The western population, which is more numerous, extends from the Point Culver area of Western Australia, westward along the southern coast then northward along the west coast to Shark Bay, Western Australia, and including all offshore islands with suitable habitat.

Rock Parrakeets are mainly a coastal species but have occassionally been recorded feeding several kilometres inland. They are usually seen on cliffs and rocky areas, on sandy beaches right up to the waters edge, on sand dunes, coastal swamps, in mangroves and on tidal flats.

This species feeds on the seed of native and introduced grasses as well as the seed and fruit of shrubs and succulents that grow along the coastline. The seed of pigface which commonly grows throughout much of their range is particularly relished.

It is thought that at least some of the moisture required by this parrakeet is extracted from the succulents it feeds on, although regular drinking from fresh water soaks have been observed.

Rock Parrakeets are not easily flushed when feeding and will not fly until the last possible moment, then rise and fly some distance before settling again. During periods of strong winds, which are quite prevalent in much of their range, they are even more reluctant to fly, preferring to remain in the shelter of the low vegetation and rocks.

When this species is eventually flushed it utters a soft, two syllable, whistling call.

Feeding usually takes place along or close to the coast on the mainland as well as on offshore islands. After breeding most parrakeets leave the nesting islands and often flocks are formed, particularly while feeding and watering. On occasions these flocks may number in excess of 100 birds.

Rock Parrakeets have been observed to fly distances in excess of 20 km from offshore islands, where they roost at night, to feed on the mainland during the day and then return again to roost.

The field sighting of this species which remains most vivid in Stan's mind was also his first sighting. It occurred during the afternoon of August 24, 1971 on the beach at Coffin Bay Sanctuary, South Australia. Three Rock Parrakeets were first seen feeding on the seed and flowers of the wild pigface which grew profusely adjacent to the sandy beach. After some 15 minutes the parrakeets flew low and fast, in a manner reminiscent of the flight of a Sandpiper and landed right at the water's edge. They ran to and fro picking at the debris left behind as the small waves receeded then surged again.

The whole exercise was carried out in a totaly un-parrotlike manner and in a situation so abhorrent to that expected of a psittacine that it made Stan wonder if he was in fact actually observing a member of the parrot family.

These Rock Parrakeets ran along the beach, again in a Sandpiper-like manner, for some time before rising quickly and flying to a rocky outcrop which was dotted with low vegetation, where they settled again.

So often one's first sighting of a bird species is the one always remembered — the vision of these Rock Parrakeets scampering along the deserted beach is etched in Stan's mind for all time.

The flight of this species is fast, floating and irregular with slight sideways deviations which is characteristic of the species. Periods of flight are often intermingled with intervals of gliding. Landings are negotiated at high speeds which results in the desperate spreading of wings and fanning of tail to avoid crash landings.

Breeding in the Wild

Rock Parrakeets nest in holes or crevices in or under rocks, often in positions concealed by vegetation, as well as in sea-bird burrows. Due to the vunerability of these nesting sites to attack by predators such as rats, cats, lizards, snakes, foxes, etc., breeding occurs almost exclusively on islands, where predation is greatly reduced. There are few records of this species breeding on the mainland.

Breeding has been recorded from August to December and fresh eggs observed as late as February. Under favourable conditions double brooding may occur. Three to five white rounded oval eggs are laid on bare rock, earth, sand or guana, depending on the circumstances of the nest location. Incubation lasts approximately 18 days and the incubating female is fed at the nest site by the male.

Fledging usually occurs about 30 days after hatching. Half feathered young have been observed scampering around extensive nest sites and begging to be fed by parent birds outside the nest entrance. Juveniles usually moult into adult plumage at about four months old.

After breeding most Rock Parrakeets disperse from the rocky nesting islands to form family parties and flocks at established feeding grounds.

Aviculture

The first recorded breeding of this species was achieved by Dr. Russ in Germany prior to 1880. A pair housed in a flight cage in his birdroom nested in a nest box normally used for starlings, laid three eggs, one of which was fertile and eventually fledged a healthy youngster. Hopkinson in his *Records of Birds Bred in Captivity* 1926 provides the references, Russ, Bull 1880, page 680 and Neunzig page 752.

There are no other early recorded breedings of this species either in England or Europe. Edward Boosey of the Keston Bird Farm, England, recorded a clutch of infertile eggs prior to 1939. (*Avicultural Magazine*, 1952 page 157).

In Australia the first recorded breeding was by G. E. Pearce of Port Augusta, South Australia in 1936 for which he received the bronze medal of the Avicultural Society of South Australia.

Cayley in his *Australian Parrot* 1938 records a further breeding in 1936 by W. L. Penny of Plympton, South Australia.

Adelaide Zoo reared two young during the 1941–42 season.

Stan's early attempts to breed Rock Parrakeets, during the 1960's all ended in disasters of one kind or another. The first problem to overcome was obesity, which led to lethargic behaviour and then in turn to death. This problem was eventually overcome by reducing the maintenance diet to dry millets only plus supplementary foods of sprouted millets and plenty of greenfood, as well as providing housing with a minimum length of 3.6 m (12 ft) to promote flying exercise.

The next problem to solve was how to bring this species into breeding condition and then induce them to breed. Reduction of body weight plus a nutritious diet went a long way to solving this problem but still Stan could not breed this species.

Rockeries with numerous and varied nesting sites were constructed but the only breeding carried out in them was by hordes of mice. The next attempt to provide acceptable nesting sites was in the form of a cement wall constructed at one end of a large aviary which had a number of holes and hollows built into it. This aviary was 7.2 m (24 ft) long, 2.0 m (6 ft 6 in.) wide and 2.0 m (6 ft 6 in.) high and housed six Rocks. Over the next few years not once was a Rock Parrakeet seen to enter one of these sites.

During the next decade odd youngsters were reared by fostering eggs or young or by hand-rearing but still no parent reared young.

It was not until 1986 that Stan successfully parent-reared this species.

Three surgically sexed pairs were housed in an aviary 4.5 m (15 ft) long, 1.8 m (6 ft) wide and 2.4 m (8 ft) high which was roofed for 1.5 m (5 ft) at each end, with an open section in the middle. The back and both side walls were of solid material and the front, which faced north, was open. Both roofed sections had concrete floors and the open centre section had a natural earth floor which supported the growth of a course rush-like grass.

Nest sites in the form of hollow logs and nest boxes were hung high under the sheltered sections. The chosen sites were a vertical hollow log with a 15 cm (6 in.) internal diameter and a natural entrance spout and two horizontal nest boxes 22 cm (9 in.) long, 15 cm (6 in.) wide and 15 cm (6 in.) high with 10 cm (4 in.) entrance spouts.

The first pair to commence breeding selected a nest box and laid the first of three eggs on September 29, 1986 and the others at two-day intervals. The chicks hatched October 19, 20 and 22, 1986 after a 19-day incubation period. Fledging occurred after 36 days for the first two chicks and 38 days for the third chick.

Number two pair nested in the hollow log and laid their first egg September 30, 1986 and completed the four egg clutch at two-day intervals. This clutch was infertile. The first egg of a second clutch of three eggs was laid November 23, 1986 and completed at two-day intervals. Only the last egg hatched after a 20-day incubation period and the chick duly fledged after 37 days in the nest.

Pair number three laid only two eggs, the first on October 27, 1986 and the second four days later. The hen incubated poorly so the eggs were fostered under Blue-wings. One egg hatched and the chick eventually fledged with its Blue-wing nest mates.

During the following two years young were reared each season from this colony yet we did not feel the full breeding potential of each pair had been reached. Although two clutches and sometimes three clutches of eggs were laid each season by each hen, only a maximum of one nest of young was reared by each pair per year.

We believe that in most cases of colony breeding, regardless of the species of parrot, better results would be obtained on average, over a number of seasons, if the pairs were housed individually. The stress created by the dominant birds, on themselves and the rest of the colony, reduces optimum breeding results. Of course there are occassional exceptions to the rule.

In an effort to prove this point, in 1989 two pairs from this colony were housed individually in fully roofed aviaries with concrete floors, which measured 3.6 m (12 ft) long, 1.2 m (4 ft) wide and 2.1 m (7 ft) high. The aviaries were adjacent to each other and the front half of the partition wall was wire mesh so that the pairs could see each other yet still feel secure in their own territory (aviary), thus eliminating all stress. No fighting took place between the pairs through the wire partitions.

Each pair was provided with horizontal nest boxes 25 cm (10 in.) long, 15 cm (6 in.) wide and 15 cm (6 in.) high with 10 cm (4 in.) entrance spouts. (See Figure 6E).

Number one pair laid their first of four eggs August 31, 1989. Three chicks hatched and duly fledged at the end of October. The first egg of the second clutch of four was laid November 14, 1989 again three chicks hatched and eventually fledged on January 12, 1990.

Number two pair laid their first egg of a clutch of five on August 19, 1989. Five chicks were hatched but two died in the nest and the remaining three fledged on October 7, 1989. The first egg of a second clutch of five was laid on October 21, 1989 but the eggs in this fertile clutch failed to hatch.

These results, although not excellent, were an improvement on the results obtained when housed on the colony system.

Jim's experiences were similar when birds were housed in a colony. In a 5 m (16 ft) long by 2 m (6 ft 6 in.) wide by 2 m (6 ft 6 in.) high aviary with the back half enclosed and the roof at the front half being wire, fostering was required to maximise production. Adult mortality was also at unacceptable levels.

Once the birds were moved to totally roofed aviaries and one pair per aviary production increased and mortality was reduced. In Jim's semi-rural environment, owls and goshawks are a constant problem and Rock Parrakeets seem to suffer the ill effects of the raptor scare tactics to a greater degree than other species in the collection.

John Raymond of Sydney, New South Wales, who had good results with Rock Parrakeets housed in suspended wire breeding aviaries kindly supplied us with the following details:

The aviaries measured 75 cm (2 ft 6 in.) wide, 90 cm (3 ft) high by 1.8 m (6 ft) long and were suspended 90 cm (3 ft) above ground level. These aviaries were situated side by side under the heavy canopy of a eucalypt tree which provided them with shade for much of the day all year round, and thus a relatively cool environment. John's property backed on to what he termed a "humid gully" which he believes, when added to the shade, created a cool, moist environment particularly suited to Rock Parrakeets.

The birds were housed one pair to an aviary and provided with John's preferred "african lovebird" nestbox in which both sides of the box protrude six or seven centimetres beyond the front of the box and a horizontal shelf extends between these protruding sides just below the entrance hole which is near the top of the box. (See Figure 6D).

John's diet consisted of one part Japanese millet, one part French white millet and one part of equally mixed sprouted sunflower and canary seeds together with a few sprouted mung beans, silverbeet, dockweed and milk thistle in season. Grit was included in the seed mix and cuttlebone supplied periodically.

Between one and four youngsters were reared per nest with an average of two per nest recorded over a number of years. John's Rocks double brooded on only a few occassions during this period.

Rock Parrakeets are often double brooded and we have had pairs lay three clutches of eggs but have never had them rear three nests of young.

Recently John McCrory of Melbourne, Victoria, informed us that during the 1989–90 season a pair of Rock Parrakeets held in his aviaries nested three times and reared three clutches of four healthy youngsters.

A vertically hung hollow log 75 cm (2 ft 6 in.) long with an internal diameter of 17 cm (7 in.) with a natural entrance spout facing downwards and situated near the top was the chosen nest site.

The first clutch of four eggs was laid in September, three young hatched and duly fledged in early November. Four more eggs were laid prior to the young leaving the nest. Three were hatched and eventually fledged in December. After approximately six weeks break during which time the pair moulted four more eggs were laid in early February. All hatched and then fledged during March. This is the only record of Rock Parrakeets rearing three clutches in the one season that we are aware of.

Precocious Young

Rock Parrakeets have remarkably precocious young from a very early age.

Bill Schwarzenberg of Victoria observed the following precocious behaviour of young Rock Parrakeets bred in a colony situation in his aviaries during the 1989 season.

The colony was provided with similar vertical nesting logs which measured 60 cm (2 ft) long with an internal diameter of 10 cm (4 in.) and quite rough interiors. Each log had an entrance hole near the top and an inspection access near the bottom.

Three pairs in the colony had nested and Bill had monitored the nestlings progress. When the most advanced nest of young were at the early pin-feather stage, Bill was amazed to find the small nesting chamber empty, where there had been three young the previous day. As he had opened the inspection door he had thought he heard scuffling inside the log, so he felt up and down the hollow but to no avail. He then peered through the entrance hole towards the top of the log and there, clinging to the rough sides of the vertical log were the three tiny nestlings, barely at the pin-feather stage.

Bill, now being alerted to this strange behaviour, closely observed the other nests of young, all of which in turn reacted similarly to his inspections.

Stan also experienced this precociousness with the young of a pair housed in an aviary with ground frequenting birds. Dead tussock grass and a number of wire tunnels covered with grass had been placed over the floor of the aviary shelter as cover for the co-occupants. The Rocks were never seen to frequent this ground cover.

The three young in the nest box were half feathered when a 40°C day temperature was encountered. When Stan made one of his many checks of young in the nests, as is his policy on extreme days, all the young Rocks had abandoned the nest box. Eventually, after a considerable search, the chicks were located huddled together in the grass at the end of a tunnel. When the temperature dropped towards evening, the young were replaced in the nest box where they remained until fledging two weeks later.

This is the only species of Neophema, or Australian parrakeet for that matter, where the young return voluntarily to the nesting site after fledging. This practice sometimes proves disruptive to pairs intent on having another clutch.

In explanation of the precocious behaviour of young Rock Parrakeets, we would suggest that almost all ground nesting birds have precocious young, which enables them to relocate as a defence mechanism against predation, nest damage or adverse weather conditions.

Sexing

The sexing of adult Rock Parrakeets can often be difficult, in fact, so difficult that we never attempt to pair birds of unknown sex for breeding. Surgical sexing is recommended for this species.

If visual sexing is the only means available, we submit the following information as a guide only. Males usually appear to be slightly larger with brighter colouring and a broader, brighter blue frontal band. The sex of some individuals appears quite obvious yet the results of Jim's surgical sexing often proves the reverse.

Immatures are even more difficult to sex but may be surgically sexed successfully a few weeks after fledging.

Display

The male advances towards the female in an erect stance with his head bobbing while uttering a soft twittering whistle. The female will often respond by begging for food and courtship feeding usually follows.

Nests

Rock Parrakeets have been recorded nesting in a variety of nest sites and situations including constructed rockeries, under objects on aviary floors, in nest boxes of various designs and in numerous types of hollow logs.

We feel that due to the readiness of this species to adapt to such an array of nest sites, the consideration should be for a nesting situation that is easily managed. Obviously nest boxes must be the popular choice above rockeries, holes under water dishes or hollow logs, etc., for they are relatively vermin free and are easy to construct, clean and inspect.

Our preferred nest box is a horizontal box 28 cm (11 in.) long, 18 cm (7 in.) wide, 20 cm (8 in.) high and fitted with an 8 cm (3 in.) entrance spout, filled with fine wood shavings to a depth of 2.5 cm (1 in.). This parrakeet shows a marked preference for nest boxes with entrance spouts (see Figure 6E). Because Rocks often breed on into the summer months it is advisable to drill a few 6 mm (¼ in.) ventilation holes in opposing sides of the box to help lower the temperature within.

All nest sites should be positioned out of direct sunlight and under shelter.

Nesting and Hatching

Female Rock Parrakeets usually spend several days in and out of the nest site prior to the first egg being laid.

We have recorded eggs laid as early as August 19 and as late as February 16, which is an extensive breeding period particularly considering the information was recorded in the harsh climatic conditions of the south-western suburbs of Sydney. The majority of eggs were laid during September, October and November.

A Newly hatched chicks. Note, dense long down on the young of this species.

B Nine-day-old nestlings. Note the dark grey down.

C Twenty-one-day-old nestlings.

D Thirty-day-old nestlings almost ready to fledge.

E Cinnamon.

F Cinnamon Yellow Rock Parrakeet. Photographed in the wild.

Plate 17. Rock Parrakeet — nestlings and mutation.

165

The usual clutch of three to five white, rounded oval eggs are laid at two and occasionally three-day intervals. We have consistently recorded 19 and 20-day incubation periods but never the oft quoted 18-day period.

Chicks hatch with a relatively long, dense silvery grey down all over the upper body and head but sparse on the underparts. At five days the down colour changes to light grey, the eyes have started to open at eight days and at nine days old early pin-feather development can be seen. Tail and flight feather development is prominent at 13 days and by 16 days old the nestlings are plump, fluffy, downy chicks with pin-feathers coming through the mid-grey down. At three weeks old the tail and flight feathers are well developed but the body is still heavily downed. When four weeks old the chicks are almost fully feathered.

We have recorded fledging periods of 35 to 39 days which are longer than all other Neophemas. The young fledge as duller versions of the adults with no sign of the blue wash on the face, very little or no blue frontal band, yellow beaks and large black eyes which lack the iris rings.

By ten weeks old the beak has turned brown, the blue frontal band is visible, there is a slight trace of pale yellow lores but there is no sign of the blue wash on the face. Adult plumage is attained at four to six months old.

Mutations

CINNAMON

The only occurrence of a distinct mutation in this species, that we were aware of, was that of a sighting which was related to Bernard Enders, late of Victoria, of a yellow bird in a small wild flock in South Australia, many years ago. Apparently unsuccessful attempts were made to capture this yellow mutant. Years later Bernard obtained some Rock Parrakeets said to be descendants of normal birds taken from the same flock.

Eventually a yellow youngster which had hatched with red eyes was fledged from this stock, but unfortunately the youngster survived only a few weeks. From all reports this bird was definitely a cinnamon.

Later still Gordon Dosser of Victoria obtained two normal young Rock Parrakeets from Bernard Enders which were from the same stock and proved to be both males. One of these males, when mated to a totally unrelated female produced red-eyed young in two nests during the 1989 season.

In 1990 a red eyed chick was successfully fledged which is a typical Cinnamon in appearance and retains bright red eyes (See Plate 17E). Gordon tells us that early indications suggest this mutation is probably of recessive inheritance which could be expected in an advanced Cinnamon mutation retaining red eyes.

Apparently this mutation does not conform closely with the Cinnamon Yellow mutation fledged by Bernard Enders, which presents two possibilities. Either two distinct mutations are involved or it is a variable mutation (as some Cinnamon mutations are.)

We would like to wish Gordon the very best of luck with this Cinnamon mutation.

CINNAMON YELLOW

We are in recent receipt of a good quality colour photograph of two Rock Parrakeets, one of which has bright yellow underparts, light olive-yellow upperparts, pale fawn flight feathers and tail plus a diluted blue frontal band and a horn coloured beak.

The photograph, which is of wild birds, was sent to Barry Hutchins of South Australia about 10 years ago. The mutant depicted is a typical cinnamon yellow. (See Plate 17E).

PIED

A Pied mutation has been worked on for many years by John Pace of Victoria.

Stan visited John some years ago and was able to view both live and mounted specimens. The yellow pied spots and blotches occur in varying degrees over the body as well as on the flight feathers.

John has had considerable difficulty with this recessive mutation and states it is still far from established.

Hybrids

Hybrids have been recorded between this species and the Elegant Parrakeet, the Blue-winged Parrakeet and the Orange-Bellied Parrakeet (*Records of Parrots Bred in Captivity*, Arthur. A. Prestivich, London 1954).

Bill Schwarzenberg of Victoria bred hybrids from a Rock Parrakeet and a Turquoise Parrakeet during 1990.

BOURKE'S PARRAKEET

Plate 18. Bourke's Parrakeets (pair) male on right.

BOURKE'S PARRAKEET

Neopsephotus (Neophema) bourkii

Derivation:
Neopsephotus:
   Neo — from *Neos,*
     Greek for new
   Psephotus Greek for "inlaid
     with pebbles"
bourkii — after Sir Richard
   Bourke, Governor of
   New South Wales 1831–37

23° 27'

Description:

   Length 21 cm average
   Weight 48 g average
     (females usually lighter
     than males)
   Appearance as in Plate 18

Classification

Controversy has existed for decades amongst ornithologists and aviculturists alike, as to whether the Bourke's Parrakeet really belonged in the Neophema genus.

Various differences between this parrakeet and the other members of the Neophema genus have been cited, such as behavioural patterns, body structure and colour, as well as the fact that there is no record of fertile eggs having been produced with any other Neophema species.

These physical and behavioural differences have now been reinforced by the findings of the research program "Biochemical Systematics of Parrots" being carried out jointly by the Museum of Victoria, C.S.I.R.O. and the Australian National University, which establishes the relationship between species and genera by blood analysis.

The results of this research proves beyond doubt the Bourke's Parrakeet *(Neopsephotus bourkii)* has no close relationship with the Neophema group and in fact belongs in an isolated monotypic genus as does the Budgerigar, *Melopsittacus undulatus* and the Cockatiel, *Nymphicus hollandicus.*

Gregory Mathews introduced the monotypic genus *Neopsephotus* to accommodate the Bourke's Parrakeet in his *A List of the Birds of Australia* page 137, 1913, when he recognised its dissimilarities to other members of the genus Euphema (Neophema) as it was then known.

Mathews separated the subspecies *Neopsephotus bourkii pallida* from the Musgrave Ranges, Central Australia which he found to be paler in colour; *Austral Avian Record,* Volume 3, page 57, 1916. This subspecies is not currently recognised.

Earliest Report

The first published report of the Bourke's Parrakeet is in Mitchell's *Three Expeditions into the Interior of Eastern Australia,* Volume 1, page 18, 1838, under the name *Nanodes bourkii.* Mitchell had discovered this species on the banks of the Bogan River in New South Wales. This species was first described and illustrated by John Gould in his *Birds of Australia* Volume 5, Plate 43, 1841, as *Euphema bourkii.*

In 1891 Salvadori in his *Catalogue of Birds in the British Museum,* Volume 20, page 570, relocated this species into the Neophema genus as *Neophema bourkii.*

Range, Habitat and Field Notes

The Bourke's Parrakeet's movements throughout its vast arid and semi-arid range appear totally unpredictable. There are periodic eruptions from the usual barren habitat into more arable regions, often far from their normal distribution.

These movements must in part be influenced by seasonal conditions and hence food supply.

Determining the exact range of a species which inhabits such vast, remote regions of predominately inland Australia is impossible. Basically the Bourke's Parrakeet is found in a broad belt which extends across inland Australia as far east as the Narrandera, Tottenham, Trangie and Bourke regions of western New South Wales and the Cunnarnulla, Charleville districts of south-west Queensland. Then north-west to the Mt. Strangway area (which is 60 km north of Alice Springs) in the Northern Territory and further north to peak at about Mt. Wallaston.

From the eastern extremities the range extends as far south as the northern tip of Spencer Gulf then across the western deserts to the arid coastal region of Western Australia from the vicinity of Shark Bay to Exmouth Gulf.

The field records gathered by the Royal Australasian Ornithologist's Union for *The Atlas of Australian Birds* suggests the possibility of eastern and western populations which are divided by a gap in the range between 139°E and 140°E. This gap coincides with what may be natural barriers namely the Flinders Ranges in the south and the Lake Eyre region to the north. Bear in mind this 1° longitudal break in the recorded range represents a distance of 100 km (63 miles) which would hardly act as a barrier particularly during cool conditions or when the desert blooms after rain.

The southern limits of this species' range appears to be restricted by the Nullarbor Plains.

This species favoured habitat is dry acacia scrub with a predominance of mulga or open mulga woodlands, although sightings have often been made in open sandy, stoney, and sand ridge country with low, sparsely scattered scrub, particularly *acacia*.

Unlike its co-habitor of these arid regions of inland Australia, the Scarlet-chested Parrakeet, *Neophema splendida,* the Bourke's Parrakeet's existence in a locality appears to be dependant on the availability of water. There are numerous records of this species drinking from a variety of water supplies, often before dawn or after dusk and sometimes hours after dark. This unusual drinking habit has probably evolved as protection against predator attack.

The congregation of large flocks near water during periods of drought are also regularly reported. There is also a lack of any records of this species obtaining moisture from desert plants. All of these points suggest the inability of this species to survive beyond a reasonable flying distance from a water supply during hot weather conditions.

Their diet consists of the seeds of various native grasses, shrubs and herbaceous plants as well as vegetable matter. This species is particularly fond of the seeds of various acacias which are eaten from the shrub and foraged for on the ground.

Field sightings are usually of small groups when drinking, feeding on the ground or roosting in shrubs close to the ground. They are usually quiet and allow close approach before fluttering a short distance to perch or resume feeding. Large flocks have been observed during drought periods in the vicinity of watering places.

The Bourke's Parrakeet is increasing in numbers throughout its range, probably due to the continually increasing establishment of permanent water supplies for grazing stock throughout the arid interior.

Although this species is in relatively good numbers in the arid and semi-arid interior of Australia, Stan encountered extreme difficulties in sighting this parrakeet. For over 30 years he ventured into the outback whenever possible, often into the range of the Bourke's Parrakeet yet was never rewarded with a sighting. Then in recent years excursions were taken to known drinking places of this species, purely to achieve a sighting yet all attempts were thwarted by heavy falls of rain. This either eliminated the bird's need to drink at a particular place, flooded the access roads to the desired locality or just dampened the enthusiasm for bird watching trips.

To add salt to the wound, Stan's son Ray, who has only a passing interest in birds, drove to Bourke in western New South Wales for a long weekend and sighted a pair of Bourke's feeding on the roadside just a few kilometres from the town.

In June 1990 an expedition was launched into the Great Victoria Desert in South Australia for the express purpose of sighting the Bourke's Parrakeet and the Scarlet-chested Parrakeet. Stan and his wife Jill were members of this party.

The group entered the desert west of Coober Pedy with seven people and all the necessary equipment packed into two four-wheel drive vehicles.

During the mid-morning of June 16th, 1990 on the first day's travel into the desert the first sighting of the Bourke's Parrakeet was made. While travelling along the four-wheel drive track known as the Ann Beadell Highway at a point about 50 km west of Mabel Creek Station, Belinda Gillies-Isles saw three small parrots, which she suspected were Bourke's Parrakeets leave the track just in front of the Landcruiser.

They flew some 300 m across the open, stoney terrain to a broad, dry, shallow depression which was heavily studded with acacia scrub about 3 m (10 ft) high. As the group moved through the scrub, Bill Schwarzenberg startled a flock of about a dozen Bourke's which uttered their musical twitter as they flew off, low to the ground in an attempt to avoid the cold, blustering, south-west winds which prevailed.

After moving another 200 m through the scrub 16 Bourke's were sighted sheltering from the wind in two low branches of a bushy acacia, perched shoulder to shoulder, no more than a metre above the ground. They allowed Stan and his friends to approach to within 20 m before flying off rapidly, calling as they went, to vanish into the bushes some distance away. Sex differences could be distinguished with the aid of binoculars during this sighting.

Another four were noted in nearby bushes and a larger group a little further away. It was estimated that approximately 50 birds were sighted in the area.

These Bourke's Parrakeets displayed none of the quiet disposition which allows close approach, described by other observers. This was probably because of the appalling cold, windy conditions which made life unpleasant for man and birds.

This sighting was made within flying distance of permanent water. A further 800 km were travelled across this totally waterless desert and then on to the Nullarbor Plains without another Bourke's Parrakeet being sighted.

The flight of this species is fast yet floating and usually close to the ground. Stan observed this species in fast flight through acacia scrub which was negotiated with ease and in preference to flying a metre or two higher for clear uninterrupted flight. He also observed flight across open country which was fast and relatively low, about 3 m (10 ft) above the ground. The rapid, short wing beats were interrupted by brief moments of gliding.

At the commencement of flight a noticeable "whirring" of the wings can be heard above the whistling, twittering calls.

Breeding in the Wild

Breeding in the wild has been recorded from August to December but with this species whose arid habitat is so extensive, breeding must be influenced to some extent by weather conditions.

Three to six white, rounded oval eggs are laid on a base of rotten wood in hollows in fallen timber, stumps and limbs and trunks of small trees. Nest sites have been recorded from just above ground level to 3 m (10 ft) high.

Incubation is carried out by the female for a suggested period of 18 days. The young are fed by both parents and fledging occurs about four weeks after hatching. They leave the nest similar in colour to the adult female and are fed by the parents for a further two weeks while foraging in a family group. At four to five months old the youngsters moult into adult plumage.

Loose colony breeding has been observed in mallee woodlands with some nest sites only 6 m (20 ft) apart.

Double brooding during good seasons seems certain when records of nest sites being used twice in the same season are considered.

175

Aviculture

The Bourke's Parrakeet is the most successful avicultural subject of this group and amongst the most freely bred psittacine in the world.

Hopkinson in his *Records of Birds Bred in Captivity* 1926, reports the first breeding was by Dr. Carl Russ of Germany in 1880 (*teste Russ in Bull*, 1880 page 680).

In Great Britain the first breeding was by Mr. Fasey who reared five young in 1906, an achievement which earned him the Avicultural Society Medal. (*Avicultural Magazine* 1906, pages 276 and 343).

Early avicultural records of this species in Australia appear to be confined to two references in Alfred J. North's *Nest and Eggs of Birds Found Breeding in Australia and Tasmania* Volume 3, pages 154 and 155, 1912. They are as follows;

Percy Peir of Marrickville, Sydney, New South Wales, in correspondence to the Australian Museum states that in 1904 he received five pair of Bourke's Parrakeets from Adelaide, South Australia. He found them timid at first but they adapted well to aviary life, yet at the time of writing (October 5, 1909) they had still not bred.

Dr. W. Macgillivray of Broken Hill, New South Wales, stated he had two pairs of Bourke's Parrakeets in his aviary which he had procured from a bird-catcher who found them breeding near the Queensland border in the summer of 1902–03. He had hopes they would breed.

It is obvious this species was held in aviaries in Australia during the early part of this century and it seems unlikely such a prolific aviary bird would not have bred, yet there are no records until 1927 when Dr. W. Macgillivray of Broken Hill, New South Wales, again writes;

"These birds are easily bred in captivity, but each pair should be kept by themselves, and the fledged young should be separated as soon as the female shows signs of nesting again, as the male is apt to persecute the young and cause them serious injury. One pair that I have in an aviary reared five broods in the one season extending from August until the middle of March." (The Charming Bourke Parrot by Dr. W. Macgillivray, *The Emu*, Volume 27, page 67).

Dr. W. Macgillivray's breedings of the Bourke's Parrakeet prior to or during 1926 are the first recorded breedings of this species in Australia, that we are aware of.

176

The first breeding in South Australia was achieved by Mr. Simon Harvey of Adelaide in 1930.

Stan's first breeding of this species was in 1958 from a pair housed in an aviary 1.8 m (6 ft) long, 1.8 m (6 ft) wide and 2 m (6 ft 6 in.) high which was fully roofed and had a sand filled floor. The nest was a hollow log 15 cm (6 in.) in diameter and 40 cm (16 in.) long and hung in a partially inclined position. Four eggs were laid at two-day intervals, three chicks were hatched and all were fledged.

We have bred this species in almost every conceivable type of housing from the recommended small, enclosed, fully roofed aviaries with concrete floors to large planted aviaries, in suspended wire aviaries as well as small breeding cabinets.

The Bourke's Parrakeet will inflict only negligible damage to shrubs etc., in planted aviaries and are totally reliable when housed with finches, doves, softbills and in most instances other Neophemas, who are more likely to be the aggressors.

Successful colony breeding has often been recorded although more often than not colony breeding of this species is unsuccessful. The dominant pair will often stress the remainder of the colony, resulting in only the dominant pair breeding. No colony can be deemed successful unless the majority of pairs breed.

We have successfully flock bred this species when housed with finches, doves, softbills and with individual pairs of King Parrots, Crimson Wings, Regent Parrots, Superb Parrots, Princess Parrots, Scarlet-chested Parrakeets, Turquoise Parrakeets, Elegant Parrakeets, Blue-winged Parrakeets and Cockatiels. It is advisable to keep a close watch for any sign of aggression from the larger parrots.

The following interesting results of a particularly prolific pair of Bourke's Parrakeets which were housed in a breeding cabinet 1.3 m (4 ft 6 in.) long, 60 cm (2 ft) high and 60 cm (2 ft) deep and situated in a wire mesh fronted birdroom, was recorded by Herbert (Stud) Baker of Sydney, New South Wales.

The 12-month-old, untried pair were placed in the cabinet in May 1984 and provided with a hollow log 15 cm (6 in.) in diameter, 30 cm (12 in.) long, with entrance gained down the top, which was hung inside the cabinet at a 45° angle.

This table provides the breeding results from the first clutch until the time of writing;

Clutch No.	No. of Young Fledged	Month and Year
1	1	July 1984
2	2	October 1984
3	2	December 1984
4	5	February 1985
5	4	April 1985
6	3	July 1985
7	4	September 1985
8	3	November 1985
9	3	January 1986
10	2	March 1986
11	5	June 1986
12	3	September 1986
13	5	November 1986
14	3	January 1987
15	2	March 1987
16	2	July 1987
17	4	September 1987
18	4	November 1987
19	2	February 1988
20	Nest destroyed by mice	April 1988
21	2	July 1988
22	3	September 1988
23	Infertile eggs	November 1988
24	3	January 1989
25	4	March 1989
26	1	May 1989
27	2	August 1989
28	4	November 1989
29	3	February 1990
30	Infertile eggs	May 1990
31	3	August 1990
32	3	October 1990

The statistics provided by these records are astounding. This pair of Bourkes have nested 32 consecutive times in six years and three months, reared young on 29 occassions, and had two clutches of infertile eggs as well as one nest destroyed by mice.

Eighty-seven young have been fledged during every month of the year, which is an average of three young per nest over the 29 fertile clutches. The maximum period between clutches was four months.

We have examined the youngsters reared on many occasions and all have been of exceptional quality.

This pair of "super" Bourkes are still in good condition at the time of writing, having just fledged another three young — October 1990.

We feel these incredible results can be directly attributed to the expert husbandry provided by this 85-year-old aviculturist. Good work Stud.

Sexing

Sexing of adult Bourke's Parrakeets is obvious — the male has a blue frontal band and the female does not, although very occassionally an adult female will carry an odd blue feather on the brow. The female also has generally duller plumage and the narrower pink margins of the breast feathers give a more scaled appearance on the chest.

Immatures, who are similar to adult females only duller, are difficult to sex although males often have a larger head with a more pronounced brow. Sexing by the white underwing bar is unreliable.

Display

The male advances towards the female, stretched to his maximum height, with tail fanned slightly and wings extended from the body at the shoulders to display the blue shoulderband, while uttering a low "chirruping whistle"-like call.

Nests

This species will accept any reasonable nesting facility and have been bred in every variety of nest box and hollow log possible.

As with other prolific members of this group we are inclined to use commercially available nest boxes, either vertical or horizontal, and discard them after a year or two. Boxes are preferable to hollow logs because of the ease of accessibility and cleaning.

A pair that refuses to breed in a nest box can often be stimulated to breed by providing them with a hollow log. We prefer a hollow log, with a removeable lid for access, about 30 cm (1 ft) long, an internal diameter of 15 cm (6 in.), a 2.5 cm (2 in.) entrance hole near the top and hung either vertically or partially inclined.

Any of the recommended nest fillings are suitable for this species. Bourke's nests are often subject to an excessive build up of droppings, particularly when large clutches are being reared, thus necessitating nest cleaning during the rearing period.

Nesting and Hatching

The Bourke's Parrakeets breeding season normally extends from late July until March, although in areas with a reasonable climate, breeding may occur throughout the year.

In the relatively mild climate of the Sydney region of New South Wales this species has been recorded to lay eggs during every month of the year.

A few weeks prior to breeding the pair will frequent the chosen nest site. Three to six, but usually four or five white, rounded oval eggs are laid at two-day and occasionally three-day intervals. Incubation usually commences with the laying of the second or third egg, and usually extends for a period of 18 to 20 days. In September 1991 we recorded the remarkable and inexplicable incubation period of 13 days for each of the first three eggs of a six egg clutch which was laid by a closely monitored pair. Our most commonly recorded incubation period is 20 days.

The chicks hatch with a fluffy, pale greyish white down and the eyes open at seven days old. Early pin feather development is visible at 12 days and is well advanced at 15 days old. Chicks are half feathered when 20 days old and fully feathered and ready to fledge soon after. We have recorded fledging periods of 23 to 34 days.

Youngsters leave the nest similar to adult females with black eyes, which lack the iris rings, and a yellow beak.

Parent Bourkes usually show no sign of aggression towards their fledged young, even while rearing their next clutch, but the young may be removed from their parents with safety three weeks after fledging.

Immature Bourke's Parrakeets usually moult into adult plumage at four to six months old. Sexual maturity is usually reached about 12 months old although both sexes of this species have been known to breed at six months old.

Mutations

Australian Primary Mutations

THE CINNAMON RANGE

TYPICAL SEX-LINKED CINNAMON

In 1962 Herbert (Stud) Baker of Sydney, New South Wales, bred from a normal pair, what we would term a typical sex-linked cinnamon Bourke. The chick hatched with deep red   es which appeared almost normal in

A Brooding hen with newly hatched chicks.

B Eight-day-old nestlings.

C Fifteen-day-old nestlings.

D Twenty-eight-day-old nestlings.

E Blue-rumped Opaline female.

F Pink Lacewing female.

Plate 19. Bourke's Parrakeet — nestlings and mutations.

181

A Recessive Cinnamon
(Normal eye, Diluted feet).

B Recessive Cinnamon (Red-eyed) female.

C Cream (Cinnamon Cream) pair.

D Fallow (slight dilution and red-eye).

E Rose (Opaline) female.

F Pink (Opaline Cream) female.

Plate 20. Bourke's Parrakeet — mutations.

colour by the time it fledged. The bird, which proved to be a female, had the typical flesh coloured feet and horn coloured beak of a Cinnamon and appeared identical to the sex-linked Cinnamon mutation established in America.

Unfortunately this mutation was not established from this source, although what appears to be the same mutation is currently being developed elsewhere in Australia.

RECESSIVE CINNAMON (Normal eye) See Plate No. 20A

A Recessive Cinnamon mutation was bred in Tony Pine's aviaries in the Hunter Valley region of New South Wales during the 1989 season. It is a male which hatched with normal coloured eyes, has dark fawn wing markings, typical diluted feet and beak, and was produced from a supposed pair of Split Cream Bourkes.

Development of this new mutation is now in progress.

RECESSIVE CINNAMON (Red eyed) See Plate No. 20B

Another distinct recessive pallid Cinnamon mutation which hatches with red eyes and retains them at maturity, as most very advanced Cinnamons do, appeared in the Cessnock area of New South Wales during the late 1970's. This mutation reappeared in Brian Anderson's aviaries in Singleton, New South Wales during 1983.

Currently this mutation is being developed by Gary Hall of Moree.

CINNAMON YELLOW

A Cinnamon Yellow type mutation which was described as a straw or biscuit colour with all the normal pink, very diluted blue areas, red eyes, flesh coloured feet and horn coloured beak, was bred by David Judd of Mildura, Victoria in 1978. It was a female but died before reproducing, hence the genetic inheritance is unknown.

CREAM (Cinnamon Cream) See Plate No. 20C

This mutation is in fact an advanced Cinnamon (i.e., close to lutinoism) that hatches with, and retains red eyes and is of recessive inheritance, as most advanced Cinnamon mutations are.

The Cream Bourke is the counterpart of the Cinnamon Yellow mutation of green based birds and perhaps "Cinnamon Cream Bourke" is a more descriptive and accurate terminology.

There is a marked difference between the sexes in this mutation, the female in most cases being much yellower then the male, who is often just a little yellower than a Cinnamon. Enormous variation also occurs between individuals of the same sex.

Extreme difficulties have been encountered by many breeders of this mutation due to some pairs failing to breed true to the laws of genetic inheritance. Cream pairs sometimes produce normal as well as Cream chicks while some of the normal young produced from a Cream to split mating don't appear to be splits. Just to confuse the issue, other breeders have not experienced these problems with Cream Bourkes.

The only logical explanation for these apparently impossible results is for two distinct mutations to be involved. If the two mutations concerned are both of recessive inheritance it is possible for some mutant birds to be split for the second mutation and vice versa, and for others not to be split at all. Likewise some splits may be split for either one of the mutations involved while others may be split for both.

It is also possible that the second mutation involved is of a sex-linked inheritance. If this is the case female "Creams" could either be just recessive "Creams" or a secondary mutation combining both "Cream" and the second mutation, which could well be a sex-linked Cinnamon. This could explain why most Cream males are more like Cinnamons, because perhaps in fact that is what they really are — Cinnamons split for Cream, while the rare Cream coloured males are a combination of both mutations.

The two mutation theory explains the failure of so many "Cream to split" and "split to split" matings to produce "Creams" and also why many breeders have bred Cinnamons from Cream stock.

The origin of this mutation in Australia points to one person, Phillip Irwin of Spears Point, New South Wal .

Phil first bred a Cream Bourke in 1968 from stock carrying excessive blue which he was line breeding in the hope of breeding a blue Bourke. Unfortunately this first Cream chick, which had red eyes and pink feet, was lost before fledging.

It took the next ten years to breed five Cream birds, all of which were cocks. In 1981 the first Cream hen was bred. A predominance of cocks prevail in this stock, Phil averages four Cream cocks to every Cream hen bred.

Phil has also experienced the production of normal coloured young from Cream pairs but has never produced any other type of mutant from his Cream stock which seems to nullify the theory of the involvement of a second mutation in explanation of this mutation's unpredictable breeding results.

The best coloured and most successful strain of Cream Bourkes we have seen was developed by Daryl Gray of Singleton, New South Wales from a single Cream bird which was used to produce two distinct blood lines by outcrossing with two unrelated normal Bourkes.

This strain of "Creams" has always bred true to the recessive laws of inheritance, probably because the only birds used in the development were the original Cream and two normals, and significantly no other mutant birds.

FALLOW See Plate No. 20D

The Fallow mutation is yet another of the Cinnamon group which has a general slight dilution of the normal colouring, bright red eyes which are retained at maturity, flesh coloured feet and legs with diluted claws and beak usually retaining a brownish wash. This is normally a recessive inheritance.

We have examined a few mutant birds that fit into this category, one is being termed an "Isabel" which is an ill-defined term needing clarification on an international level as to just what constitutes an Isabel mutation. Until such clarification is forthcoming, it may be wise to treat all these mutations as "Fallows".

At least two, possibly three, distinct Fallow mutations exist in Australia but to our knowledge none have been fully established as yet.

It is important to retain these closely allied mutations, distinct, one from the other.

The best Fallow Bourke Stan has seen was bred in South Australia during the 1989 season from a normal pair and is now in the care of Reg Collyer of Adelaide, South Australia, who has great plans for its future. (See Plate No. 20D)

Another variety of Fallow Bourke which has less dilution of the body colour yet retains the red eye, is being developed in Sydney, New South Wales.

DILUTE YELLOW (Black eyed Yellow)

A very yellow Bourke's Parrakeet retaining normal coloured eyes and all blue colouring in a diluted form, a pink chest, and which developed a diluted blue brow signifying a male, was bred by Keith Carter of Victoria in 1972. (Source; Peitre Vroegrop, Warrigal, Victoria).

Unfortunately this beautiful recessive mutation has not been established.

185

BLUE-RUMPED

This recessive mutation which has a blue rump and extended blue areas on the head and wings has appeared in Europe but this mutation was first recorded in a wild caught male bird which was imported by Percy Peir of Marrickville, Sydney, New South Wales from South Australia in 1904. The body of this mutant was donated to the Australian Museum in 1909 and reference is made to the study skin of this bird by Alfred J. North in his *Nest and Eggs of Birds Found Breeding in Australia and Tasmania* 1912 Volume 3, page 153.

ROSE (Opaline) See Plate No. 20E

This beautiful and often variable mutation is in fact an Opaline, and is of sex-linked inheritance as all Opaline mutations appear to be.

The exact origin of this mutation in Australia is unclear but Stan was shown Rose Bourke's incubating eggs in Joe Mattinson's aviaries in Wollongong, New South Wales during 1985. This mutation is now reasonably well established in Australia.

Sexing this mutation can be difficult — there appears to be at least two distinct strains. In one strain the male does not acquire a blue frontal band on maturity whereas the males of the other strain gain a reduced, blue frontal band on their juvenile moult. To confuse the issue even further some adult females will carry a few blue feathers on the brow. Sexing by head size is often necessary — males usually have a larger head and more pronounced brow. Surgical sexing is advisable with difficult birds.

PINK LACEWING See Plate No. 19F

A female Bourke's Parakeet with extended and accentuated pink on the belly and chest and the cream borders on the wing coverts replaced with pink has recently come into our possession.

This bird has undergone a complete moult while in Stan's aviaries without change of colour thus indicating it could be a new mutation. Verification will depend on future breeding results.

Australian Secondary Mutations

The production of an array of secondary mutations using combinations of any of the preceding primary mutations is now only a matter of time.

A combination of the Rose mutation with any of the Cinnamon range offers an exciting potential for new colours.

PINK See Plate No. 20F

Pink Bourke is the name being applied to the stunning secondary mutation combining the Rose and Cream mutations. In other words the Pink mutation is an Opaline Cinnamon Yellow.

Basically in this mutation the Cinnamon Yellow influence reduces the rose colour of the Rose Bourke to pink while the brown wing feathers are changed to cinnamon and cream, and the red eye is retained.

Many Australian aviculturists are currently working on this mutation. At the present time good specimens are rare but with outcrossing and selective breeding the strains will improve.

BLUE-RUMPED OPALINE See Plate No. 19E

A new and unexpected mutation appeared in Reg Collyer's aviaries in Adelaide, South Australia, during the 1990 breeding season, that we feel is best described as a Blue-rumped Opaline.

Reg has a strain of normal Bourkes which carry excessive blue colouring, particularly around the head. Although not Blue-rumped Bourkes, they are perhaps the forerunner of this mutation. This strain was used to produce split Rose Bourkes.

The Blue-rumped Opaline was bred from a split Rose (Opaline) cock which was normal in colour, but was from the excessive blue strain, and a normal hen which perhaps also carried the excessive blue blood. The result, the Opaline version of a Blue-rumped Bourke.

To reproduce a secondary mutation such as this the genetic potential in the parents when mated together, to produce both mutations must exist (in this case Opaline and Blue-rumped). When the genetic potential for both mutations occurs in the one chick the secondary mutation appears.

Overseas Primary Mutations

THE CINNAMON RANGE

SEX-LINKED CINNAMON

This mutation was well established in Southern California, U.S.A. when Stan toured the region as a member of an Australian avicultural group attending the San Diego Avicultural Convention in 1981.

Breeders of this mutation informed Stan that the chicks hatch with deep red eyes which darken before fledging.

There is a marked resemblance between this mutation and the sex-linked Cinnamon currently being developed in Australia.

Of course as the name implies this mutation is of sex-linked inheritance.

RED-EYED SEX-LINKED CINNAMON

Another sex-linked Cinnamon mutation which retains a dark red eye after fledging and has diluted cinnamon colouring has been developed in Europe. Apparently little interest is shown in this mutation.

RECESSIVE CINNAMON

Stan also observed a Recessive Cinnamon mutation whilst visiting Southern California in 1981. This mutation was quite similar in appearance to the sex-linked Cinnamon but he was assured by breeders that its genetic inheritance was recessive. Confirmation as to whether the chicks of this mutation hatch with red or normal coloured eyes could not be verified.

One breeder Stan spoke to at that time believed the two Cinnamon mutations in the U.S. had been unknowingly bred together resulting in considerable difficulties in sorting out each individual's genetics.

YELLOW

The Yellow Bourke is yet another of the "Cinnamon type" mutations so prevalent in this species. It appears to be a similar mutation to the variable Australian "Cream" mutation.

This recessive mutation which retains the red eye after fledging was probably the first mutation to occur in this species in Europe.

It originated in the aviaries of Mr. J. van de Brink of the Netherlands in 1957 when two Yellow young were reared from a normal pair. A total of seven Yellow youngsters with red eyes were reared from this pair during that season.

Simultaneously in 1957, Mr. Lievens of Belgium reported breeding Yellow Bourke's Parrakeets.

As with the Cream Bourke in Australia, females of the "Yellow" mutation are much yellower than males. Young males develop a light blue frontal band after their first moult.

FALLOW

This recessive mutation which is another of the Cinnamon group, apparently appeared in Europe during the development of the Yellow Bourke, although its precise origin is unknown.

The European Fallow Bourke is similar in appearance to the least diluted Fallow mutation currently in Australia.

ISABEL

The European Isabel mutation of the Bourke's Parrakeet is also of the Cinnamon group that has more dilution than a Fallow and pronounced cream scalloping on the wings. This sex-linked mutation hatches with red eyes which are retained after fledging and has only slight dilution of the beak, feet and claws.

It was first bred by G. Demarest of Holland, probably in 1959.

European authorities disagree on what constitutes an Isabel Bourke. One authority states this mutation has slight dilution of the feet, greyish beak and claws, while another indicates pink feet and horn coloured beak and claws. Both agree it retains red eyes after adulthood.

LUTINO

There is no mention of the existence of a Lutino Bourke mutation in recent European literature, but a Lutino Bourke's Parrakeet is featured in a promotional video made of J. Postema's aviaries and birds in Holland. This video has been made to advertise the opening of Mr. Postema's collection to the public.

The bird's colouring is cream, yellow, pink and white. We would assume that if the bird is a true Lutino which it appears to be on the video, it would be a sex-linked inheritance.

ROSE (Opaline)

This spectacular mutation was developed in Europe. It was first bred in 1970 by Mr. G. Goossens of Schinveld in Holland from a visually normal male and a normal female which had been bred from an Isabel cock and a normal female. In the first nest this pair produced four normal coloured cocks and three Rose hens, with a total of four Rose hens being reared in the first season.

At first this mutation was thought to be of recessive inheritance but eventually it was found to be sex-linked.

Stan first saw living specimens of this mutation at the huge breeding complexes of "Bird Behavioural Studies" near Los Angeles in Southern California, U.S.A. in 1981. The stock had been imported from Europe the previous year.

BLUE-RUMPED

As the name implies this recessive mutation has a blue rump and usually extended blue on the head and wings which sometimes covers the top of the head and half of the wings.

Splits to this mutation can frequently be identified by blue feathers in the rump.

Secondary Overseas Mutations

YELLOW-ROSE

When a Yellow Bourke is bred with a Rose Bourke, the offspring, when mated together in a combination with the potential to produce both "Yellow" and "Rose", are also capable of producing Yellow-Rose young.

This mutation is in fact the Opaline version of the Cinnamon Yellow mutation in which the depth of rose colour is reduced, the wing marking becomes cinnamon and pale yellow, the beak and feet are diluted in colour, and a red eye is retained at maturity.

Sometimes this mutation is referred to as a Pink Bourke in Europe.

ISABEL-ROSE

This secondary mutation is the result of the second generation breeding of the Isabel and Rose mutations. Perhaps it is best described as the Isabel form of the Rose Bourke.

It is similar in colour to the Rose Bourke with wing markings the same as the Isabel, diluted beak and feet colouring and red eyes.

FALLOW-ROSE

Fallow-Rose Bourkes are produced when a Fallow and Rose are bred together and the resulting splits are mated together in a combination capable of producing both mutations. Such a mating can also produce the Fallow form of the Rose Bourke.

It is similar to the Rose Bourke with cinnamon wing markings, diluted beak and feet colouring, red eyes and is considered the most beautiful of the "Rose" secondary mutations.

Hybrids

There is no authentic record of a hybrid being produced from the Bourke's Parrakeet with any other species.

190

Some years ago a hybrid was reported between this species and a Scarlet-chested Parrakeet in the U.S. Photographs accompanying the report showed the hybrid to be identical to an immature Bourke's Parrakeet. This so-called hybrid was obviously the result of this species, as well as other Australian psittacine desert species, ability to withold fertilised eggs within the body for many months until conditions suitable for breeding prevail. The hen Bourke, who was housed with a male Scarlet-chested, had lost her mate three months prior to laying her clutch of eggs. Obviously the fertile clutch of eggs was retained by the hen.

This phenomenon has also been recorded on a number of occasions in Princess Parrots.

Many years ago Stan produced numerous clutches of infertile eggs from a female Bourke's Parrakeet and a male Budgerigar.

REFERENCES

J. M. Forshaw — *Australian Parrots* 2nd Edition, Lansdowne Press.

RAOU — *The Atlas of Australian Birds*. Melbourne University Press.

A. Lendon — Neville Cayley's, *Australian Parrots in Field and Aviary*. Angus and Robertson.

Australian Aviculture — Avicultural Society of Australia, C/o G. Hyde, 52 Harris St, Ellminyt Victoria 3249.

Bird Keeping in Australia — Avicultural Society of South Australia, P.O. Box 4, Blackwood, South Australia 5051.

The Emu — Royal Australian Ornithological Union, 21 Gladstone St, Moonee Ponds, Victoria 3039.

The Australian Birdwatcher — Journal of the Bird Observers Club of Australia, P.O. Box 185, Nunawading, Victoria 3131.

H. P. M. Zomer — *Grass Parrakeets and their Colour Mutations*. Gwendo — Arnhem.

B. R. Hutchins and R. H. Lovell — *Australian Parrots,* A Field and Aviary Study. Avicultural Society of Australia, 1985.